QUALITY PLANNING FOR THE LIFE SCIENCE RESEARCHER

Meeting Quality Assurance Requirements

QUALITY PLANNING FOR THE LIFE SCIENCE RESEARCHER

Meeting Quality Assurance Requirements

Amy Cross-Smiecinski
Linda D. Stetzenbach

CRC Press
Boca Raton   Ann Arbor   London   Tokyo

Library of Congress Cataloging-in-Publication Data

Cross-Smiecinski, Amy
    Quality planning for the life science researcher: meeting quality assurance requirements/
written by Amy Cross-Smiecinski and Linda D. Stetzenbach; illustrated by Shirley Burns.
     p.   cm.
    Includes bibliographical references and index.
    ISBN 0-8493-8693-4
    1. Life sciences—Research—Quality control. I. Stetzenbach, Linda D. II. University
of Nevada, Las Vegas. Harry Reid Center for Environmental Studies. III. Title.
QH315.C735 1994
574′.072—dc20

                93-47331
                CIP

No claim to original U.S. Government works
International Standard Book Number 0-8493-8693-4
Library of Congress Card Number 93-47331
Printed in the United States of America 1 2 3 4 5 6 7 8 9 0
Printed on acid-free paper

DEDICATIONS

Amy Cross-Smiecinski dedicates the book:

> To my parents, Jack and Jane Cross, my husband, Ralph, and our children, Sarah and Peter, for their love and support.

Linda Stetzenbach dedicates the book:

> To Klaus and our daughters, Erika and Kristen, for their love, patience, and understanding.

PREFACE

This text has been prepared by a university quality assurance (QA) specialist who has worked as a biologist and chemical analyst for the U.S. Environmental Protection Agency (U.S. EPA) and its contractors, and a research microbiologist with experience in clinical, academic, research, and U.S. EPA arenas. Both have also worked as cooperators with the U.S. Department of Energy (U.S. DOE).

The authors use their experience in these diverse settings along with the associated respect for life science researchers attempting to meet QA requirements to create a useful and relevant guide to help the biologist document and relate his/her quality assurance program to current QA practices.

Previous QA guidance documents have been geared to the chemical and physical sciences with little focus on environmental problems involving biological science. This document, however, has been written as a QA primer for research biologists with QA requirements described and their usefulness and relevance to the life sciences discussed.

The purpose of this book is to assist the biological researcher in the development of a quality assurance plan that meets the requirements required by many sponsoring agencies. This book is to be used as a guidance document and is not to serve as a template by the research scientist. It is designed to provide a logical connection between policy requirements and the re-search scientist focusing on quality assurance in a life science study. It is intended to assist the scientist in achieving an understanding of QA and successfully implementing a quality assurance plan.

An issue that is not and cannot be resolved in this document is the standardization of quality assurance and quality control terminology. The authors use U.S. EPA-referenced definitions and terminology, where possible. All other terms and defini-tions are borrowed from John Taylor,[1] the U.S. DOE, and other authorities. Each term is presented in a glossary at the back of the text and also as a footnote on the page of text where the term first appears.

This book has three basic objectives:

1. To help the research scientist label and arrange study com-ponents into a practical QA plan for a research program;
2. To help the researcher build plan-specific quality goals and the means to attain them into the QA plan; and
3. To help the life scientist appreciate the service that quality assurance provides.

It is not the intent of the authors to "mandate" quality assurance requirements, or to make the scientist feel mistrusted. This text is to be used as a handbook, with explanations and definitions of QA elements for all parties involved in a quality assured program and, therefore, it should be helpful to re-search scientists, nontechnical managers, and reviewers of bi-ology projects.

AUTHORS

Amy Cross-Smiecinski, M.S., is a Quality Assurance Specialist at the Harry Reid Center for Environmental Studies, University of Nevada, Las Vegas.

Mrs. Cross-Smiecinski received her B.S. degree from New Mexico Highlands University, Las Vegas, New Mexico in 1974. She received her M.S. degree from the same university while working with the U.S. Environmental Protection Agency on organic pollutant pathways research. She has held positions as biological researcher and chemical analyst, and provided quality assurance support to cooperative agreements with the U.S. EPA and the U.S. Department of Energy. She served as coordinator of an international quality assurance materials program for the Human Exposure Assessment Location Program with the World Health Organization.

Mrs. Cross-Smiecinski is a member of the Energy and Environmental Division of the American Society of Quality Control. She has authored several journal articles and made presentations at national conferences. Her current major interests are in implementing multiagency QA requirements within a multidiscipline research center setting.

Linda D. Stetzenbach, Ph.D., is the Director of the Microbiology Division at the Harry Reid Center for Environmental Studies, University of Nevada, Las Vegas. Dr. Stetzenbach graduated in 1970 from the University of Arizona, Tuscon with a B.S. degree in microbiology. She worked as a clinical microbiologist for 10 years before returning to the University of Arizona where she obtained an M.S. degree in 1984 and a Ph.D. degree in 1986 from the Department of Microbiology and Immunology. Before assuming her present position, she served as postdoctoral fellow in the Department of Veterinary Science, University of Arizona, where she developed and licensed a monoclonal antibody for recognition of *Giardia* cysts.

Dr. Stetzenbach is a member of the American Society for Microbiology, the Mycological Society of America, the American Society for Testing and Materials, and is a registered microbiologist with the American Society of Clinical Pathologists. She has received an Award of Excellence from the U.S. Environmental Protection Agency, Environmental Monitoring Systems Laboratory — Las Vegas, in recognition for participation in the U.S. EPA bioremediation research project in Valdez, Alaska following the Exxon Valdez oil spill in Prince William Sound and in U.S. EPA indoor air quality monitoring survey. She has received research funding from the U.S. EPA, the U.S. Department of Energy, industry, private sector, and local governments.

Dr. Stetzenbach is the author or co-author of several book chapters and journal articles and has presented numerous invited lectures at national and international meetings. Her current research interests include methods development to enhance identification and enumeration of microorganisms, indoor air quality, water quality, and bioremediation.

TABLE OF CONTENTS

Introduction 1
     QA Plans 3
     Planning Research — The QA Project Plan 5
     Formatting 6
     Document Control 7
     QA Plan Revisions 10
     References 12
     The QA Plan as a Tool 13
     Seventeen QA Plan Elements 13

Element Descriptions 17

I.   Title Page 19

II.   Table of Contents 23

III. Project Description 27

IV. Project Organization and Responsibilities 29

V.   Quality Objectives 33

VI. Site Selection and Sampling Procedures 45

VII. Sample Custody 49

VIII. Calibration Procedures and Frequency 53

IX. Measurement Methods 55

X. Data Reduction, Verification, Validation,
and Reporting 57

XI. Computer Software 63

XII. Quality Control Requirements 67

XIII. Audits and Assessments 77

XIV. Preventive Maintenance 81

XV. Calculation of Data Quality Indicators 85

XVI. Corrective Action 89

XVII. QA/QC Reports to Management 91

Glossary 93

References 99

Index 103

Introduction

Quality assurance (QA)[a] is a management philosophy developed to assure the reliability of data. It provides a means to integrate planning, assessment, and interpretation of study data. Quality assurance is a means to substantiate the validity of research data and provide a basis for supporting regulatory decisions resulting from that data. Quality control (QC)[b] is a means to document standards of performance in the collection of data. These terms are often used incorrectly or are misused interchangeably.

Quality assurance principles and practices have previously been published, and guidance documents have been written to assist the researcher in preparation of QA plans.[2-4] These publications, however, were written with the emphasis on chemical

---

[a] QUALITY ASSURANCE (QA): The total integrated management program for assuring the reliability of data. A system for integrating the quality planning, quality assessment, and quality improvement efforts to meet user requirements.

---

[b] QUALITY CONTROL (QC): The routine application of procedures for obtaining prescribed standards of performance in the measurement process.

laboratory procedures. They do not address many of the QA issues unique to biological research projects. Sampling, measurement, and QC methods for biological parameters can become more complex than for chemical parameters due to concerns for the viability of the organisms under study and because life science measurements are often subjective rather than standardized. The difficulties of applying QA to life sciences are summed up in a statement by Miller[5] regarding the challenges facing microbiology researchers when confronting QA concepts. He states, "Microbiology was founded as a science of observation and interpretation in which standard deviations and tolerance limits are foreign." The research scientist, however, must be assured that the data generated were obtained in accordance with good laboratory practice and approved standards.

Quality programs in the life sciences were found formerly only in the medical and clinical microbiology fields. Quality control in the medical microbiology laboratory, for example, was mandated in 1967 with the passage of the Clinical Laboratory Improvement Act demonstrating that QC is an essential component of quality health care.[6] With this act, QC measures covering all aspects of clinical laboratory testing were implemented. While the basic QA and QC philosophy described for the clinical microbiology laboratory has application to the research laboratory, numerous aspects of the guidance are difficult to apply to non-clinical microbiological, biotechnological, and other fields of life science research.

A QA program can require 10 to 20% of an investigator's time and generally 15% of total analysis time.[7] While preparation of a QA plan may initially present a formidable challenge to the research life scientist, the investment in effort ultimately is proven to be well spent. This is particularly evident when the quality of the results is shown to be absolutely unquestionable, when study costs must be estimated, when every aspect of the study has a documented explanation, when a statistician does not have to be consulted on an emergency basis at the end of the study, and when any detail of the study can be reviewed and explained long after its completion.

The composition of a QA plan provides the opportunity to chart the course of a study from costs to results and subsequent work, but many researchers do not know where to start when viewing a list of sponsor's QA requirements.

QA PLANS

There are basically two kinds of QA plans. The first type is usually called the quality management plan (QMP),[c] although the terms QA requirements document or QA program plan are occasionally used. The quality management plan specifies the quality policies instituted from the corporate or management level on down to the studies level (Figure 1). The other type of QA plan is usually called the quality assurance project plan (QAP)[d] (Figure 1) and indicates the QA and QC requirements at the project level. This project-level document is often subordinate to the quality management plan. That is, it must be written to meet the requirements in the upper-tier document. A given laboratory or research program has one quality management plan, but has a QA project plan for each of its research projects within the overall program.

Most governmental and other agencies that fund scientific research require in their QA policy that studies awarded funding have a QA plan. As an example, the U.S. EPA requires that a written and approved QA project plan be signed and operational prior to the collection of any data. The purpose of that requirement is to ensure that the money invested in the project will yield scientifically valid conclusions related to a principle hypothesis and that the research is documented from the planning stage through the conclusion.

For many QA programs, there is a three- to four-level tier of documents (Figure 1). These include the QA plans, the standard operating procedures (SOPs),[e] and the scientific notebooks.

---

[c] QUALITY MANAGEMENT PLAN (QMP): An orderly assemblage of management policies, objectives, principles, and general procedures by which an agency or laboratory outlines how it intends to produce data of known and accepted quality.

[d] QUALITY ASSURANCE PROJECT PLAN (QAP): An orderly assemblage of detailed and specific procedures that delineates how data of known and accepted quality are produced for a specific project or study.

[e] STANDARD OPERATING PROCEDURE (SOP): A written document that details an operation, analysis, or action whose mechanisms are thoroughly prescribed and that is commonly accepted as the method for performing certain routine or repetitive tasks.

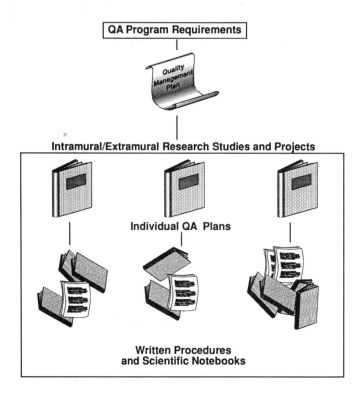

FIGURE 1. The hierarchy of QA documents.

Project SOPs are developed to meet the requirements of the QA plan and are a stepwise description of a routine task. Procedures may be common to more than one study if they apply to a task performed according to manufacturer's instructions (e.g., operation of an autoclave) or a method used throughout the laboratory facility (e.g., scientific notebook recordkeeping). Often this type of procedure requires little or no innovation on the part of the project leader other than to include it in the QA plan and document that the staff performs the procedures as directed. However, procedures may be written for a single assay (e.g., the preparation of an experimental culture medium).

Scientific notebooks can be the most important form of recordkeeping for a research study. Prizes and patents have been won and lost by the use or misuse of scientific notebooks.[8] Because they are so important to research, it is advisable to obtain a good text on the subject and possibly compose a procedure to assure that the usefulness of this method of recordkeeping is maximized. Generally, the more basic the research, the greater the use of scientific notebooks.

The authors have chosen to discuss the QA project plan part of the QA program in this book since, as mentioned previously, this document is often required shortly after award of funding; the requirements may be difficult to understand, and little or no guidance exists for the life scientist in this area.

PLANNING RESEARCH — THE QA PROJECT PLAN

In obtaining funding and approval to carry out a study, it is difficult for the researcher to resist the urge to jump immediately into data collection. Lack of formal planning, however, often causes a scramble with the statistician or a series of laboratory procedure repetitions just prior to the final report's deadline. While good planning does not guarantee that problems will not occur, it can minimize them, eliminate the embarrassing ones, save money, prevent overdue products, and even enhance project image. The QA plan, often thought of as a bureaucratic requirement, can actually be a valuable planning tool.

The basic sequence followed in a scientific investigation incorporates the following seven steps:

1. Define data quality objectives (DQOs)[f] based on allowable errors.

---

[f] DATA QUALITY OBJECTIVES (DQOs): Statements describing the level of uncertainty a decision maker is willing to accept in results derived from environmental data. DQOs are defined when the results are going to be used in a regulatory or programmatic decision (e.g., deciding that a new regulation is needed, setting or revising a standard, or determining compliance).[9] DATA QUALITY: Features and characteristics of data that bear on their ability to satisfy a given purpose. The characteristics of major importance are accuracy, precision, completeness, representativeness, and comparability.[2] These characteristics are defined later in the text and in the glossary at the end of the document. Data quality can also be expanded to include timeliness and value to the client (e.g., Can the results be applied to the clients' needs?).

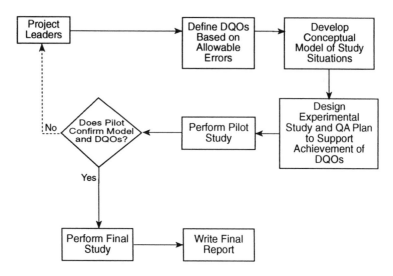

FIGURE 2. The basic sequence of steps followed in a scientific
investigation.

2. Develop conceptual model of study situations.
3. Design study and QA plan to support DQOs.
4. Perform pilot study.
5. Revise study based on pilot or redefine the DQOs.
6. Perform final study.
7. Prepare final report.

A basic guide for planning a study with these events is
presented in Figure 2. This scheme incorporates QA philosophy and still allows the investigator a developmental period
or pilot study.

FORMATTING

Formatting of the QA plan should be consistent throughout
the document and meet the requirements of the organization
and/or sponsor. Header styles, print style, justification, pagi-
nation, and other formatting of figures and tables vary throughout this book to express the general flexibility of QA plan
requirements.

```
┌─────────────────────────────────────┐
│                                     │
│   Document: _____         │
│   Revision Number: _____       │
│   Date: _____            │
│   Page: _____ of _____      │
│                                     │
└─────────────────────────────────────┘
```

FIGURE 3. A document control plate.

DOCUMENT CONTROL

Often QA programs require the use of document control. Document control is a procedure used primarily to assure that study personnel are using the most current version of the QA plan and other documents that must be kept current such as written procedures.

Document control can be accomplished in a number of ways ranging from simple to complex depending on the sponsor's requirements, details of the study, and the number and proximity of personnel or organizations involved with the study. Strict document control procedures may be necessary in a case where the research study results are to be presented in court or if they will have a bearing on a regulatory action (e.g., licensing by the Nuclear Regulatory Commission). Most small research studies need a simple document control system that assures that personnel are not using outdated documents. Large research projects that have many documents to control and personnel in several facilities will use a more detailed document control system. This will prevent any document or researcher from being overlooked during revision distribution time.

The controlled status of a QA plan can be indicated using a document control plate[3] located in the right upper corner of each page of the document. The plate contains revision information so that it can be determined at a glance if a researcher has the most recent document. A common form of document control plate for a QA plan is shown in Figure 3. In this figure, the term "Document" refers to an abbreviated QA plan title or other identifier. The "Revision Number" should start at 0 for the first unrevised QA plan and be increased 1, 2, 3, etc. with each subsequent revision of the document. "Date" is the date of the composition or revision of the document. Figures 4 and 5 show practical examples of document control plates.

Controlled documents can also be indicated using a document control stamp (Figure 6) of a particular color ink usually red. The red stamp prevents any duplicated copies of the QA plans from being confused with the controlled copies. In Figure 6, the number shown on the stamp is a unique assigned number to help locate the controlled copies. In cases where it is necessary to retrieve the old versions of the QA plan, the stamp numbers of the retrieved copies are inventoried to determine what copies are missing. This is especially helpful in a study where the QA plan is evolving quickly and it is critical that researchers have the current version in use. In addition, a list of personnel receiving the document should be maintained for reference when distributing revisions.

Figure 7 illustrates a document control list for the QA plan of a small research project. For this study example, the document control list is adequate to assure the current status of the researchers' QA plans, and a document control stamp would be unnecessary. Figure 8 shows a document control record for a large study that involves personnel located in several facilities. Such a record is associated with a detailed document control procedure that is necessary to track large numbers of QA plan revisions and other materials such as the study's written procedures. These sign-off sheets are updated every time there is a revision in the QA plan.

Figure 9 illustrates a document control agreement form that may be used to assure that personnel have read, understood, and agreed to use a particular document and subsequent revisions. The agreement is signed by each staff person before beginning to use the document. This procedure may be viewed as demeaning and is not usually used in a research setting. It is generally used for studies with extreme requirements.

Document: Indoor Air QA Plan
Revision Number: 3
Date: 02-15-91
Page: 14 of 28

FIGURE 4. An example of a QA plan document control plate.

Document: Balance X-1 SOP
Revision Number: 0
Date: 09/24/90
Section 1
Page: 2 of 5

FIGURE 5. Another practical example of a document control plate.

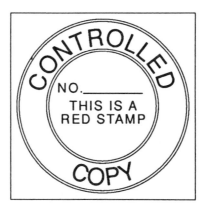

FIGURE 6. A document control stamp.

DOCUMENT CONTROL LIST

Document _Desert Tortoise Survey Quality Assurance Project Plan_

ORIGINAL/REVISIONS DISTRIBUTED

DOCUMENT HOLDERS	0:3-14-89	1:4-23-90		
Rodney Smith	✔ AJS*	✔ AJS		
Sarah Jones	✔ AJS	✔ AJS		
Matt Pedersen	✔ AJS	_No longer on study_		
Ella Peck	✔ AJS	✔ AJS		
Merkie Holloway	✔ AJS	_New to study_ ✔ AJS		

*initials of person performing document distribution   4/15/93

FIGURE 7. A document control list for a small research project.

QA PLAN REVISIONS

Often problems are encountered, and details emerge that were not discussed in the original QA plan. A revision of the plan is made, and using a simple method of document control, a revision summary page is included between the cover page

University and Community College System of Nevada
YUCCA MOUNTAIN SITE CHARACTERIZATION PROJECT
DOCUMENT TRANSMITTAL/ACKNOWLEDGEMENT RECORD

TO: FROM: UCCSN YMSCP
                                                                      Document Control Center
                                                                      Harry Reid Center for Environmental Studies
                                                                      4505 Maryland Parkway
                                                                      Las Vegas, NV 89154-4009
                                                                      (702) 895-3382

                                                             TRANSMITTAL DATE:

DOCUMENT TITLE:

DOCUMENT REVISION:

                                        DIRECTIONS

Sign/date in black ink below to confirm that the
above directions have been followed, and return
this transmittal record, with the obsolete material,
as appropriate, to the above address by: _____
                                                                      RECIPIENT

_____   _____
          DUE DATE   DATE

                                                                                          4/6/93

FIGURE 8. A document control record for a large multi-organization
              research project.

and the table of contents. This revision summary lists the text
changes that were made and the dates of the original QA plan
and its revisions. Revisions of the QA plan are distributed to
all personnel listed on the title page[2] and at the end of the table
of contents. When listing personnel, position titles such as

**AGREEMENT TO COMPLY WITH QUALITY ASSURANCE PLAN
FOR THE BIOTECHNOLOGY TEAM**

My signature below indicates that:

    1. I have read the QA plan.

    2. I have read the standard operating procedures that are unique to my research area.

    3. I accept the responsibility of adhering to the procedures outlined in this QA document.

_____   _____

          Signature   Date

FIGURE 9. A document control agreement form.[10]

laboratory technician or QA specialist may be preferable to individual names on long-term studies.

Revisions of the QA plan and other pertinent documents such as written procedures are usually the responsibility of the project leader, but this task and its frequency can be explicit in the QA plan if it is not addressed elsewhere.

REFERENCES

In addition to document control of the original and revised QA plan, specific references to primary methods, procedures, validation studies, and supporting information for the research are also required. These may be placed in a separate reference section, cited in the text, or presented as footnotes. If a reference is not available, a copy should be attached to the QA plan. References include the usual information:

- Author
- Reference title
- Source, date, edition
- Volume, page, year
- Document number or other identifiers

THE QA PLAN AS A TOOL

In composing the QA plan, the project leader considers the document's users: research personnel, QA staff, management, and others who are technically oriented in specific areas. The QA plan should be written so that a technical person who is unfamiliar with the research can understand the document. The QA plan must be useful. This may require maximizing the flexibility of the QA requirements.

The QA plan is intended to be flexible and can be constructed to fulfill only the minimum QA plan requirements, or it can contain other topics the researcher finds important, in addition to the associated procedures, sampling plans, and other details to result in an all-encompassing research document. The all-encompassing document tends to reduce iterations and review cycles and aids in final report preparation.

SEVENTEEN QA PLAN ELEMENTS

The 17 elements described in this book are intended to be a basis for the document and not to hamper the author's creative abilities to produce a useful and dynamic QA plan. The following is a listing of the QA plan elements most commonly required:

I.   Title page
II.   Table of contents
III.   Project description
IV.   Project organization and responsibilities
V.   Quality objectives
VI.   Site selection and sampling procedures
VII.   Sample custody
VIII. Calibration procedures and frequency
IX.   Measurement methods
X.   Data reduction, verification, validation, and reporting

XI. Computer software
XII. Quality control requirements
XIII. Audits and assessments
XIV. Preventive maintenance
XV. Calculation of data quality indicators
XVI. Corrective action
XVII. QA/QC reports to management

Each element listed is addressed in the QA plan with the addition of any other topic that the sponsor requires or the researcher chooses. The depth of detail provided by the researcher for each element is variable. An element determined to be irrelevant to the study can usually be excluded from the QA plan with a brief explanation[2] to assure the reader that the element has not been overlooked.

> *Example:* A study that uses **no instrumentation** (e.g., studies where plate count determinations are assessed by laboratory personnel without an automatic plate count reader), Elements VIII, Calibration Procedures and Frequency, and IX, Measurement Methods, could be eliminated with brief explanations.

> *Example:* For procedures for which **accuracy cannot be assessed** (e.g., current monitoring methods for airborne microorganisms indoors), within Elements V, Quality Objectives, and XV, Calculation of Data Quality Indicators, the sections on accuracy could be eliminated with a statement discussing the reasons why accuracy cannot be assessed for that particular procedure in the project.

Because of the ever-changing nature of research, more time should be spent on developing the parts of the QA plan that assure every finding is recorded, verified, and validated, rather than on those parts that assure that processes are performed the same way each time. Basically, as long as the research team has clear-cut objectives in mind and documented, the measurement system is "in control", and each step, procedure, and result are accurately recorded, verified, and validated, the study is well-planned for attaining quality.

Obviously, not all scientific investigations are alike, and some topics pertinent to any one study may not have been described previously in a sponsor's QA requirements documents. For example, Type I and Type II error determinations and the division of responsibility are two useful tools discussed in this book (Element V, Quality Objectives) that are not considered to be QA plan requirements in most programs. A selection of optional additions to the QA plan include a work plan and a safety plan.

In addition to a sponsor's QA plan requirements and optional additions, individual program QA requirements may apply. These requirements are usually specified in the project's applicable quality management plan.

Element Descriptions

Each of the following chapters describes the intent of the associated QA plan element.

I.

Title Page

The title page (Figures 10 and 11) usually contains the following items:

- Document control plate
- Title
- Name and address of the researcher or project leader
- Name and address of the sponsoring organization
- Project, contract, cooperative agreement, and/or grant identification numbers
- Sponsor's approval signatures
- Research organization's approval signatures

Additional items as required by the sponsor or the researcher's organization should also be included. For many sponsors, a letter to the research project leader expressing concurrence with the QA plan replaces the approval signature. Some sponsors do not require an approval on a QA plan for externally funded research. In this case, the QA plan receives review and approval within the researcher's organization. Title page signatures for internal review usually consist of the QA staff and the researcher's superior (Figure 11).

	Document: Bot Spec QAP
	Revision: 3
	Date: 10-1-90
	Page: i of xii

QUALITY ASSURANCE PROJECT PLAN
for
Aerobiological Sampler Efficiencies Research Project

Cooperative Agreement No. CR-814342

This integrated assurance plan was developed to assure that all data generated by the project discussed in it are scientifically valid, representative, comparable, complete, and of known and acceptable precision and accuracy. The signatures of key project personnel below indicate concurrence with the procedures specified in the plan and a commitment to disseminate the plan to all project personnel.

Project Quality Assurance Plan Approval

_____ Date___/___/___   _____ Date___/___/___
Linda D. Stetzenbach, Ph.D. - Microbial Studies PI   Stephen C. Hern, Chief
Environmental Research Center   Exposure Monitoring Program
University of Nevada, Las Vegas   Environmental Assessment Division
          U.S. Environmental Protection Agency
          Environmental Monitoring System Laboratory - Las Vegas, NV

_____ Date___/___/___   _____ Date___/___/___
Amy Smiecinski, Quality Assurance Officer   David G. Easterly, Quality Assurance Officer
Environmental Research Center   Office of the Director
University of Nevada, Las Vegas   U.S. Environmental Protection Agency
          Environmental Monitoring System Laboratory - Las Vegas, NV

_____ Date___/___/___   _____ Date___/___/___
D.S. Barth, Ph.D. - Director   Daniel T. Heggem, Quality Assurance Manager
Environmental Research Center   Environmental Assessment Division
University of Nevada, Las Vegas   U.S. Environmental Protection Agency
          Environmental Monitoring System Laboratory - Las Vegas, NV

FIGURE 10. A title page for a U.S. EPA research study.

Document: Bot Spec QAP
Revision: 2
Date: 7-28-93
Page: i of vi

UNIVERSITY OF NEVADA, LAS VEGAS
Harry Reid Center for Environmental Studies

QUALITY ASSURANCE PLAN

Adaptation of Botanical Species
to Heavy Equipment Disturbance on Yucca Mountain, Nevada

for the Yucca Mountain Site Characterization Project Office
Carl Gertz, Director
101 Convention Center Drive
Las Vegas, Nevada 89109

Cooperative Agreement DE-FC08-90NV10872

Lynda Blair
Project Manager
Harry Reid Center for Environmental Studies
University of Nevada, Las Vegas
Las Vegas, Nevada 89154

Signatures of Approval

_____ Date ___/___/___
HRC QA Staff

_____ Date ___/___/___
HRC Project Manager

_____ Date ___/___/___
HRC Director

FIGURE 11. A title page for a study that does not require sponsor approval.

Example: Minimal **signature** requirements for a U.S. EPA-sponsored intramural (in-house) project with the QA plan approved and signed in the order shown from top to bottom.

Project leaders' immediate supervisor _____

QA specialist _____

Some sponsors require approval signatures by both the research and sponsoring organizations.

Example: Minimal **signature** requirements for a U.S. EPA-sponsored extramural project with the QA plan approved and signed in the order shown from top to bottom.

Research organization's project leader_____

Research organization's QA specialist _____

Funding organization's project leader_____

Funding organization's QA specialist_____

II.

Table of Contents

A table of contents is an orderly presentation of the material included in the QA plan. The table of contents for any QA plan often resembles a listing of QA plan requirements. Constructing a QA plan in this manner eases the review process somewhat, but the importance of the table of contents is in its usefulness to research personnel. Therefore, specific order is not usually mandatory. An example of a table of contents is illustrated in Figure 12. Because critical QA plan requirements may be located within several areas of the table of contents due to the author's format preference, a locator page (Figure 13) is useful. Appendices containing items such as procedures and forms are also listed in the table of contents.

At the end of the table of contents, all individuals who receive official copies and any revisions of the QA plan are listed.

	Document: Micro. Stud.
	Revision Number: 4
	Date: 6-22-91
	Page: ii of iii

TABLE OF CONTENTS

Notice i

Locator Page for U.S. EPA QAPP Elements ii

Section 1   **INTRODUCTION** 1-1
     1.0   Project Description 1-1
     1.1   Research Objectives 1-1
     1.2   Parameters Measured 1-2
     1.3   Experimental Design...1-2
     1.4   Timeline 1-3
     1.5   Project Personnel 1-4
     1.6   Quality Assurance Project Plan Specifications 1-4

Section 2   **QUALITY ASSURANCE OBJECTIVES** 2-1
     2.0   Overview of Quality Objectives 2-1
     2.1   Guidelines 2-1
     2-2   Quality Assurance Design Characteristics..2-2

Section 3   **SAMPLING** 3-1
     3.1   Sample Point Selection 3-1
     3.2   Sampling Periods 3-1
     3.3   Sampling Personnel 3-1
     3.4   Sampling Procedures 3-1

Section 4   **SAMPLE PREPARATION AND ANALYSIS PROTOCOLS** 4-1
     4.1   Reagents...4-1
     4.2   Antibiotic Stock Solutions 4-3
     4.3   Media 4-4
     4.4   Stock Bacteriological Cultures 4-6
     4.5   Protocols for Microscopic Analysis 4-7
     4.6   Protocols for Culture Analysis 4-9
     4.7   Protocols for Instrument Analysis 4-13

Section 5   **QUALITY ASSURANCE IMPLEMENTATION** 5-1
     5.1   Control of Data Quality 5-1
     5.2   Data Verification 5-3

Section 6   **DATA QUALITY ASSESSMENT AND REPORTING** 6-1
     6.1   Statistical Design 6-1
     6.2   Quality Assurance Reports to Management 6-2

Section 7   **DATA MANAGEMENT** 7-1
     7.1   Sample Collection Data 7-1
     7.2   Laboratory Analysis Data 7-2

Section 8   **STANDARD OPERATING PROCEDURES** ....separate document

Section 9   **REFERENCES CITED IN QAPP** 9-1

FIGURE 12.   An example of a table of contents.

		Document: MS QAP
		Revision Number: 1
		Date: 6-19-92
		Page: ii of v

LOCATOR PAGE

U.S. EPA QA Plan Element		**Location in MS QAP Document**
I	Title page	Title page
II	Table of contents	Table of contents
III	Project description 1.0	Project description
IV	Project organization and responsibility 1.5	Project personnel
V	QA objectives 2.0	Overview of quality objectives
VI	Sampling procedures 3.4	Sampling procedures
VII	Sampling custody 2.3	Custody of samples
VIII	Calibration procedures and frequency 4.7	Protocols for instrument analysis
IX	Analytical procedures 4.7	Protocols for instrument analysis
X	Data reduction, validation, and reporting 7.0	Data management
XI	Internal quality control checks 5.1	Control of data quality
XII	Performance and system audits 5.1	Control of data quality
XIII	Preventive maintenance 4.7	Protocols for instrument analysis
XIV	Specific routine procedures used to assess	
	data precision, accuracy, and completeness 6.1	Statistical design
XV	Corrective action 2.4	Corrective action
XVI	Quality assurance reports to management 6.2	Quality assurance reports to management

FIGURE 13. A locator page showing the location within the QA plan of the sponsor's quality requirements.

III.

Project Description

This QA plan element provides a general discussion of the project and usually includes the experimental design. The description may be brief, but should be sufficiently detailed so technical personnel who are unfamiliar with the research can understand it. Inclusion of flow diagrams, tables, charts, and timelines can save the researcher time in preparing this part of the QA plan. The project description, presented as much as possible in these forms, also provides information conveniently at a glance.

Issues to include in the project description are:

- General overview, background of the project
- Hypothesis to be tested, when applicable[11]
- Study purpose and objectives
- Uses and limitations of the results

Experimental design is discussed in this section. Included in the experimental design are:

- Measurements to be made
- Tables of data to be collected
- Flow diagrams of events

- Summary tables
- Charts/figures of how results will be presented
- Timeline of project activities

Other important information may also be included in the experimental design section such as organization and responsibilities, descriptions of pilot or preliminary studies, and component activities.

IV.

Project Organization and Responsibilities

The purpose of this section is to help the project leader organize personnel sufficiently to achieve the objectives of the research. Basically this may be accomplished by determining critical tasks and subtasks and then assuring that the responsibility for each is assigned.

The structural organization of the research personnel and their responsibilities are presented, and management structure of the research organization is outlined. The project leader decides the amount of detail needed in this section based on the size and complexity of the research organization. However, the QA staff is independent of management so that the risk of conflict of interest[12] is avoided.

The following should be included:

- Chart showing management structure and communication lines (Figure 14)
- Chart listing technical and QA/QC roles, responsibility, and authority (large, multi-organization study, Figure 15; small research study, Figure 16)

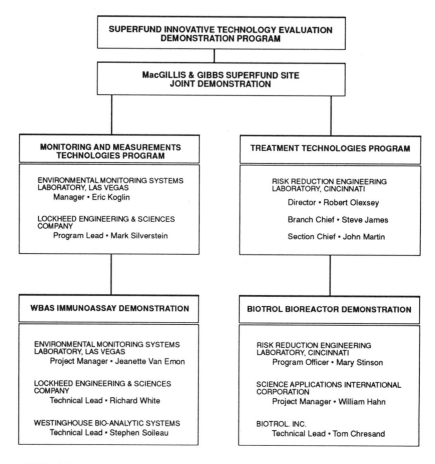

FIGURE 14. An example of a management organizational chart.[13]

Some critical tasks to consider for assignment to specified research personnel are:

- Sampling design, quality control, and operations
- Laboratory analysis and measurement design, quality control, and operations
- Document preparation, review, and control
- Data reduction, validation, and reporting
- System performance and data audit reports to management
- Health and safety
- Quality assurance commitment and program

The responsibilities of the EPA Environmental Monitoring Systems Laboratory in Las Vegas, Nevada (EMSL-LV), with assistance from its prime contractor, Lockheed Engineering & Sciences Company (LESC), for the demonstration of the WBAS immunoassay include:

- designing, overseeing, and implementing the elements of this demonstration plan;
- acquiring the necessary confirmatory data; and
- evaluating and reporting on the performance of the technology.

WBAS, the developer of the immunoassay being demonstrated, is responsible for:

- providing technical assistance to the personnel using the field kit on-site,
- performing immunoassay analyses of split samples, and
- supplying a sufficient number of field kits and laboratory reagents (for the plate immunoassay) to fully satisfy the requirements set forth in this demonstration plan.

The main function of the Risk Reduction Environmental Laboratory (RREL), through its prime contractor, Science Applications International Corporation (SAIC), Paramus, New Jersey, is to perform the BioTrol bioreactor demonstration. Aspects of the immunoassay demonstration for which RREL and SAIC are responsible include:

- providing technical assistance to the personnel using the field kit on-site,
- performing immunoassay analyses of split samples, and
- supplying a sufficient number of field kits and laboratory reagents (for the plate immunoassay) to fully satisfy the requirements set forth in this demonstration plan.

For the immunoassay demonstration, BioTrol, Inc., the developer of the bioreactor being demonstrated as a remediation technology , is responsible for:

- providing pre-demonstration test samples and
- providing technical assistance.

FIGURE 15. A description of the division of responsibilities in a large multi-organization study.[13]

Descriptions of the basic qualifications of project personnel and supplemental training needed for their assigned tasks may be included. Some sponsors require detailed training descriptions and training plans.[14] Brief resumés of personnel performing work that affects the research quality may also be referenced or supplied in an appendix to the QA plan. Any training required for the project that was completed before the start of that task is documented in this section.

Relevant certifications held by or required of individuals are pertinent information to list in this section of the QA plan. This can be written as a table listing each name with the corresponding position title, project role, tasks performed, qualifications, and supervisor.

Depending on project needs, a discussion or description of frequency and mechanisms of communications among and between researchers, contractors, the sponsoring official, and the QA staff is included. This material could be presented in a chart

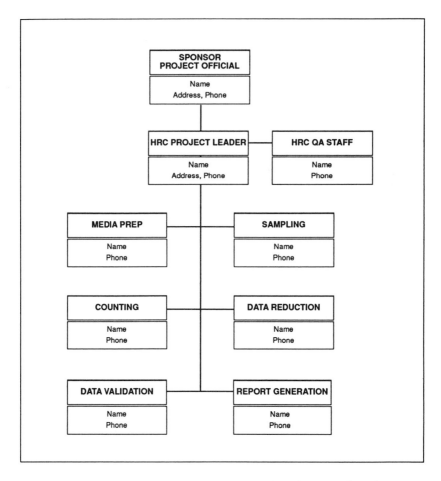

FIGURE 16.   An example of an organization chart for a small study.

that includes addresses and telephone numbers to aid communications among project personnel. Fax, cellular, and beeper numbers, if available, are also helpful. Communication numbers are especially important when team members are in different locations and/or (sub)contractors are involved in the project.

Because communication can be so important to the success of a research project, the lines of communication presented should not be described as limiting, but it should be stressed that these are minimum requirements and that free exchange of information is encouraged.

V.

Quality Objectives

In reality, scientists who produce defensible, reliable, and documented results practice QA, but they may not realize this and may not utilize commonly used QA terminology.[15] This section of the book explains the concept of quality objectives and discusses the development process and presentation in the researcher's QA plan. Because this element is often the least understood, it is also the most neglected part of many QA plans. Quality objectives are even more difficult to explain in terms of what should actually be written about them in the QA plan.

Some sponsoring organizations require all studies to be processed through a formal quality objective[9] process. This is an organization-specific phenomenon and may be determined from that organization's quality management plan. The U.S. EPA is currently developing a new process for using quality objectives to plan environmental research.[11]

Various terms (e.g., data quality objectives [DQOs], QA objectives, and measurement objectives) are encountered when discussing quality objectives. These terms are defined similarly[16] and with marginal differences.[11] These terms have been used interchangeably,[2] and the differences between them may be debatable.

- **Project or scientific objectives** are usually stated in terms of hypotheses or in the most basic research as a discussion of the research problems, nonquantitative objectives, and ideas for solutions that can be addressed by the collection of data.[11,17] This type of objective is usually presented in the Project Description section of the QA plan.
- **Quality objectives** support the study with the intention to get a desired quality of result or acceptable level of error. Quality objectives seek to ensure that the right type, amount, and quality of data are collected to resolve the research question.[11]

Some sponsors separate project or scientific objectives from quality objectives; some intermingle the terms, and others do not address quality objectives at all. The authors have noted two basic types of quality objectives, the differences between them being philosophical: data quality objectives (DQOs) and measurement quality objectives. Measurement quality objectives are specifications for a measurement system usually as a means of quality control. Data quality objectives are specifications needed for resulting project data and are used in choosing the appropriate measurement system. Therefore, measurement quality objectives are basically bench-level acceptance criteria, while DQOs are most useful in the planning stages of a study. Data quality objectives are also helpful at the end of the study when it is being determined if the resulting data were adequate to support the project and scientific objectives. See Figure 17 for an illustration of the intents and relationships of the various objectives. The researchers should remember that the type of quality objectives addressed in the QA plan depends on factors such as the sponsor's requirements and the study type and therefore its specific quality needs.

The value of quality objectives of either type is that they provide for detailed study planning maximizing the quality and usefulness of the resulting data and therefore the success of the project, while minimizing the costs and risks associated with a poorly planned study. Quality objectives are

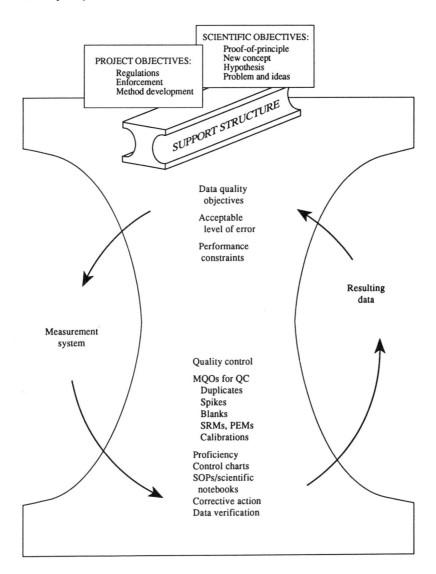

FIGURE 17. An illustration of the relationship among the objectives. The objectives, measurement system, quality controls, and resulting data all work together to achieve the project objectives.

developed and presented in QA plans in terms of precision,[g] accuracy,[h] completeness,[i] representativeness, and comparability needed to attain project objectives.[16]

> *Example:* An analogy may be drawn between **quality objectives** used in a research study and those used in a simple task. In purchasing a pH meter, characteristics such as electrode type, accessories, readout, and ruggedness in the field would be important specifications in choosing a pH meter that can serve a particular pH measurement need. Likewise, precision, accuracy, completeness, representativeness, and comparability are important specifications for measurements needed to achieve quality goals of a scientific investigation.

Quality objectives should be stated in quantitative terms as much as possible,[19] and it is important to specify the parameters to be measured and the values expected. Discussing how the quality will be determined and interpreted is also important.[19]

> *Example:* "One **quality objective** of this surface water study is to demonstrate that the population of fecal coliforms in treated sewage effluent discharged to the river is less than 1000 CFU/100 ml at a confidence level of 95%."

---

[g] PRECISION: A measure of mutual agreement among individual measurements of the same property, usually under prescribed similar conditions. Precision is best expressed in terms of the standard deviation. Various measures of precision exist depending upon the "prescribed similar conditions". In microbiology, for example, precision is the ability to isolate and identify.[18]

---

[h] ACCURACY: The degree of agreement of measurements with an accepted reference or true value. Accuracy is usually expressed as the difference between the two values, or the difference as a percentage of the reference or true value, and sometimes is expressed as a ratio. In microbiology, for example, accuracy is considered to be correct identification and confirmation.[18]

---

[i] COMPLETENESS: A measure of the amount of valid data obtained from a measurement system compared to the amount expected to be obtained under correct normal conditions.

> *Example:* "**Precision** will be measured for each sample by the use of duplicate samplers. Precision for the two determinations must be within 10% for the organism of interest."

Quality objectives are not necessary for every measurement in a study. However, measurements that are necessary to achieve the project objectives and measurements that affect the quality of the study data should conform to quality objectives. Some sponsors have specific minimum requirements. The U.S. EPA requires that, as a minimum, each QA plan addresses precision, accuracy, and completeness.[2] Yet basic research such as new-concept studies is not always amenable to these requirements.

Researchers may find it helpful to focus on the following five areas within the concept of quality objectives:

1. Determining QA objectives

    Quality objectives are stated in terms of project requirements in the QA plan, not in terms of capabilities. This is often confusing because quality objectives must be achievable by available methods, yet precision and accuracy are frequently thought of in terms of the limitations of the method used in the study. Therefore, the emphasis should be on maximizing the accomplishments of the study, while balancing this with the capabilities of available methods. Project management, researchers, and other principal project personnel may need to work closely together on the formation of objectives because once the objectives are established the measurement systems are selected to meet them. One way of planning the project is to determine the quality objectives and then determine the methods (e.g., sampling methods and analytical methods) that will support them.

    When quality objectives exceed the capabilities of available methods, the methods must be modified, or the QA plan must be revised to reflect these deficiencies. Changes in the methods used may be as simple as collecting a larger sample. However, if nonstandard or significantly modified test methods are used, the QA plan should reference laboratory validation data to demonstrate that the revised method is capable of achieving the desired performance.[12]

Example: In a project designed to monitor phylloplane bacteria, three processing techniques were compared. A stomacher blending method demonstrated significantly better recovery of bacteria from leaves than the other two procedures, and the recovery efficiency was constant over a range of sample inoculum densities.[20] The **modified method** was then used for the survey.

Research studies generally require greater flexibility in defining quality objectives and methods than other types of projects such as routine or contract laboratory projects. It is important to be realistic in developing the quality objectives and all other aspects of the QA plan.

Example: **Previous experience** with a biological treatment of pentachlorophenol (PCP) indicated that the output concentration typically varied by a factor of two under stable operating conditions with uniform feed. The **precision** of the analytical method then should be significantly less than this value or approximately ±30% to avoid contributing additional errors.[12]

2. Quantitative objectives — precision, accuracy, and completeness

Precision and accuracy are critical determinations that should not be confused. An illustration of the relationships between precision and accuracy is given in Figure 18.

Example: In establishing **precision** by using multiple aerobiological samplers at a single sampling station, quality objectives would state the acceptable standard deviation of the data derived from the multiple samplers at the level of uncertainty needed to support the scientific objectives of the study.

Example: In establishing **accuracy** with the use of molecular weight markers in gel electrophoresis, specified molecular weight markers in a control lane are used to determine the distance that a test band migrates. The quality objectives would state the acceptable distances of these markers on the gel.

Completeness needs to be described in terms of the percent of valid data required to achieve a specific statistical level of confidence.

Example: In establishing quality objectives for **completeness** in a soil sampling task, it was anticipated that 100% of 50 samples collected would be analyzed. However, during transport, the sample container

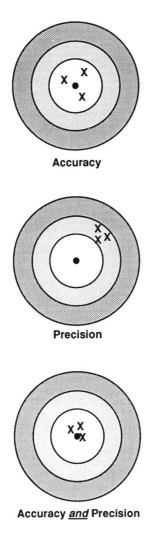

Accuracy

Precision

Accuracy _and_ Precision

FIGURE 18.   A visual display of the concepts of accuracy and preci-
               sion.[21]

caps loosened on two sample bottles, and water from the transport
cooler contaminated those samples. The samples could not be used
reducing completeness to 96%. Then in the laboratory, a technician
mishandled two more of the samples reducing the completeness to
92%.

   Quality objectives for precision, accuracy, and completeness are easily presented in a table in the QA plan. Quality

objectives for all sample or measurement methods should be included, and the units in which these quality objectives are given should be indicated. Summary tables are very helpful to the laboratory that must meet these objectives (see Figure 19 for a listing of project QA objectives).

This section of the QA plan also describes how the quality objectives are to be interpreted statistically. Often quality objectives are listed in terms of acceptance ranges. Generally, when data exceed the acceptable range, corrective action is required. However, statistical interpretations may still be possible.

Example: Individual outliers may not require corrective action if the mean value is within the acceptable range.[12]

Because the researcher defines the quality objectives, it is important to describe in this section exactly how quality objective data will be interpreted.

3. Qualitative objectives — comparability and representativeness

Comparability[j] is a parameter that is becoming increasingly important in scientific studies as it becomes apparent that many existing data sets cannot be combined to reach a valid conclusion. Comparable studies often have the same experimental protocols.

Example: To maximize the validity of data from a county-wide indoor air monitoring project, the sampling and analysis methods for each site must be **comparable**.

Example: Sampling and analysis methods of microorganisms and contaminants must be the same to **compare** pre- and post-treatment data in evaluating bioremediation of a site.

Minor differences among protocols are acceptable provided the investigators can demonstrate that the differences will not affect the comparability for that parameter.

---

[j] COMPARABILITY: The degree to which one data set can be compared to another. It is not expressed arithmetically. Comparability is achieved by using consistent methods and traceable standards.[12]

Parameter	Method	Units	Detection Range	Precision	Accuracy	Completeness
Total bacteria	AODC	cells/m^3	10^4 - ∞	± 2 s.d	90-110%	95%
	EPC	cells/m^3	10^4 - ∞	± 2 s.d	90-110%	95%
Viable vegetative bacteria	DVC	cells/m^3	10^4 - ∞	± 2 s.d	90-110%	95%
	plate count	CFU/m^3	300 - ∞	± 2 s.d	n/a	95%
Bacillus spores	AODC	cells/m^3	10^4 - ∞	± 2 s.d	90-110%	95%
	EPC	cells/m^3	10^4 - ∞	± 2 s.d	90-110%	95%
Temperature	Sensor probe	°C	0-45	± 2°	90–110%	95%
Relative humidity	Sensor probe	%RH	0-100	± 5%	95-105%	95%

AODC = acridine orange direct count
2 s.d = ± 2 standard deviations of the mean
EPC = Coulter electronic particle counter
DVC = direct viable count (Kogure method)
n/a = not applicable (no QCCS for viable counts)

FIGURE 19. An example listing of quality objectives for an aerobiological study.

Protocols, associated modifications and differences, and data usability qualifiers and limitations must be documented.

The U.S. EPA requires that, to the extent possible, all data be calculated and reported in units consistent with other organizations reporting similar data.[2] This permits comparison of data bases among organizations and related studies.

Example: When collecting data on the number of *Giardia* cysts in water, the data are commonly reported as the number of cysts per liter of water sampled, while coliforms are reported as the number per 100 ml.

Representativeness is the degree to which a sample or group of samples represents the population being studied. Generally, representativeness is expressed in terms of one parameter and is attained when a group of measurements or samples approaches the true value in that population. It is associated with collecting an adequate number of unbiased samples.

Example: Bulk leaf sampling is commonly used in field research, rather than single leaf sampling because it is considered to be a relatively quick yet **representative** process for quantifying microbial populations.[20]

The QA plan should also describe how comparability and representativeness will be achieved.

4. Other quality objectives
    Some research, by the nature of the project, requires additional types of quality objectives than those discussed above. Supplementary quality objectives should be described in detail in this section of a QA plan. These may include method detection limits and measurement quality objectives.

• Method detection limits

Example: In establishing a **method detection limit** for drop plating versus spread plates, consideration would be given to the fact that the level of detection is 10-fold less than that possible with 100-μL spread plates because drop plating uses only 10-μL of sample.[20] The quality objectives would then specify the method needed to achieve detection for the anticipated microbial concentration.

• Measurement quality objectives
• Specificity

Example: For a method to be **specific,** it must be able to isolate and detect a target microorganism. Therefore, reagents and antibiotics to control organisms other than the target microbe are added to the isolation medium.[18]

Quality objectives for method detection limits are easy to present in a table. Be sure to include quality objectives for all matrix types and to indicate the units in which these QA objectives are given. Summary tables are very helpful to the laboratory that must meet these objectives.

For some research projects, multiple measurement methods will be used and compared. A listing of quality objectives for each measurement method would be helpful. An example listing of measurement quality objectives for three methods of analysis is illustrated in Figure 20.

When measuring the same parameter for many organisms by the same method, it is usually unnecessary to list quality objectives for each organism individually. Instead, list objectives according to organism types or characteristics. In other cases where quality objectives are derived from regulations, they must be listed individually.

Performance Parameters	Field Analysis Kit	Lab. Immunoassay	GC/MS
Detection limit (ppb)	3	10 – 15	30 – 50
Linear dynamic range (ppb)	3 – 40	30 – 400	30 – 200
Accuracy	NA	NA	NA
Precision (% RSD)	10 – 15	5 – 10	21 within lab 36 between labs
Analysis time	30 minutes	2.5 hours	1 hour
# samples per analysis	5	10	1
Cost per sample	$ 30	$ 25	$ 700
Key interferents	see table 2.1	see table 2.1	no major
% matrix spike recovery	50 – 150	75 – 125	14 – 176
Field applicable	yes	yes	no

FIGURE 20. Measurement quality objectives for three methods.[13]

5. The effects of not meeting quality objectives

   Discuss the impact of not meeting one or more quality objectives in this section of the QA plan. Will the study be lost, partially useful, or salvageable with supplementary data? Is it possible that some of the project's goals will still be met? Could the confidence levels be reduced? Are there legal or regulatory ramifications?

   Determining the effect of a false positive or negative result is especially helpful in cases of regulatory or enforcement activities. Both Type I and Type II errors[k] should be

---

[k] TYPE I and TYPE II ERRORS: Tests of scientific hypotheses have been some of the most important statistical procedures used by biologists. Hypothesis testing involves evaluating a set of data and, based on an expected outcome, leads to acceptance or rejection of the hypothesis or its alternative. The rejection of a true null hypothesis is a Type I error (false positive). The acceptance of a false null hypothesis is a Type II error (false negative). Therefore, Type I and II errors express the likelihood that a false positive or a false negative will affect the study hypothesis.[22]

		Null hypothesis	
		Accepted	Rejected
Null hypothesis	true	correct decision	Type I error
	false	Type II error	correct decision

FIGURE 21.   An illustration of the relationships of Type I and Type II errors. (From Reference 22. With permission.)

considered in hypothesis testing. The relationships of the two kinds of errors and the two kinds of correct decisions to the study hypotheses are shown in Figure 21.

VI.

Site Selection and Sampling Procedures

This section describes the sampling rationale and the sampling procedures including custody, sample control, recordkeeping, and storage and shipping.[12] Depending on the project or study QA objectives, sampling can be an extremely important part of the QA plan. The selection of sampling sites and sampling strategies is based upon the QA objectives. Consequently, selection of the sampling strategy should be one that assures the achievement of the quality of data needed to reach conclusions about the experiment or decisions in regulatory/enforcement matters. Entire documents have been prepared and published[23] regarding the subject of sampling for chemical measurements, but the life science researcher encounters new challenges when sampling.

Statisticians can provide assistance in determining the number and variety of samples to be collected. They can also assist in selecting parameters that will result in clear answers to the study questions. The inclusion of statistical support in the preparation of the QA plan is, therefore, strongly suggested.

It may be helpful for the researcher to consult the following list of topics and categories concerning sampling:

- Sampling site
  If a specific site has been selected, explain why the site supports the project objectives. If the sampling location has not been determined, explain the requirements for a site as they relate to project objectives.

Example: When sampling for fecal coliforms in stream water down gradient from treated sewage discharge points, the location of the site is important. A map showing **location of each sampling site** in relation to the discharge points would serve to illustrate the rationale for site selections.

If sampling does not involve a geographic location but other criteria such as age or sex, list the requirements for selection as a sample population. These criteria will vary dependent on the research to be performed.

Example: A study where geographic **location is not critical** to selection of a sample population would be taking blood from a population based on subject age rather than geography. In this study, a map detailing the location of each sample would not be necessary.

In addition to listing the sample site requirements, the following should be included in this section of the QA plan:

- List the sample types (e.g., air, water, and soil, body fluids) and target organisms.
- List the measurement levels expected and desired in samples at the site.
- Determine the effect the site will have on data and the project objectives.
- Describe the statistical method or scientific rationale for choosing the sampling site and sampling frequencies.[12]
- Consider the factors at the site that may affect sampling[12] (e.g., road access, owner permission) and sample quality[12] (e.g., sources of contamination).
- List the measurements that must be taken in the field (e.g., pH and temperature of water samples).
- Plan for any site modifications that may be necessary (e.g., platform construction, grading) before sampling can occur.

- Sampling procedures
  Describe how samples will be collected, preserved, recorded, and shipped. The following should also be included:

  - List the measurement parameters and sample volumes to be collected for each.
  - List the sampling methods to be used (e.g., composite, grab) and reference and/or append U.S. EPA-approved or other validated standard methods. Reference and describe other, nonvalidated methods briefly and append the details.[12]
  - Describe the preparation and cleaning of sampling equipment, containers, reagents, and supplies (e.g., sampling scoop, probe, impingers).
  - Detail the calibration of field equipment (e.g., pH meter, vacuum pump).
  - Describe and reference all sample processing/handling to include the use of preservation techniques (e.g., use of formalin, azide), transportation (e.g., overnight express mail), and storage (e.g., temperature) of samples prior to laboratory handling.
  - List holding times of samples and include the maximum length of time samples can be kept before the measurement integrity is affected (e.g., microbial growth may render a sample useless after a certain length of time).
  - Describe the forms, notebooks, and procedures to be used to record sample history, sampling conditions, and analyses to be performed[2] or summarize and reference the appropriate written procedures.
  - Describe the sample control or chain-of-custody procedures[2] (Element VII of QA plan) or summarize and then reference the appropriate written procedures. Some sponsors may require a separate section for sample custody (Element VII of a QA plan).
  - Describe or summarize and reference the written QC procedures to be used to determine variability due to sampling. This includes the use of blanks and other QC sample types as well as the requirement that all sample personnel use the same procedure and are proficient, and other requirements that minimize variability.

Aerobiological sampling method: Andersen six-stage viable impactor sampler

Purpose: A forced air flow sampling method for the retrieval and size approximation of airborne microorganisms.

References: Andersen, A.A. 1958. New sampler for the collection, sizing, and enumeration of viable airborne particles. J. Bacteriol. 76:471–484.

              Seidler, R.A., and S.C. Hern. 1988 Special Report: The release of ice minus recombinant bacteria at California test sites. U.S. EPA, ERL, Corvallis.

Materials: agar-filled Petri dishes
              Andersen six-stage viable impactor sampler
              vacuum pump and Tygon tubing with in-line flow meter
              fine wire
              Kim-Wipes

Reagents: 70% ethanol

Procedure:

1)   The six stages of the sampler will be assembled with pre-labeled uncontaminated agar-filled Petri dishes and brought to the designated sampling station. The housing cover will then be removed.

2)   Tygon tubing will be attached to the vacuum port of the sampler and the vacuum pump with an in-line flow meter.

3)   The sampler will be operated at 28 liters/min. air flow for a 2, 5, or 10 minute time period determined by the research scientist according to specific needs of the experiment being conducted.

4)   At the conclusion of the experimental run the housing cover will be replaced, the tubing disconnected, and the sampler removed from the test area.

5)   The Petri dishes will be removed from the stages and incubated a time and temperature as appropriate for the media used.

6)   The number of colony forming units (CFU) will be counted and recorded on the data log sheet.

7)   The six stages will be wiped with Kim-Wipes saturated with 70% ethanol to remove any particles from the surfaces of the stages. The stages will be stacked and allowed to dry at room temperature. When dry they will be stored in the sampler.

Notes and Corrective action:

a)   The Andersen impactor sampler relies on the growth of organisms on the agar plates for detection and enumeration. The selection of agar medium and incubation conditions will influence the results.

b)   The positive hole correction chart (see section 4.6.2.1) will be used in conversion of data.

c)   Stages will be visually inspected prior to assembly with agar plates for clogged holes. If any holes are clogged they will be cleaned with a fine wire and soaked in 70% ethanol and dried prior to use.

FIGURE 22.   An example of a brief written procedure — a sampling standard operating procedure.

Written procedures are an excellent way to fulfill the above requirements, and procedures may be supplied as an appendix in a QA plan. See Figure 22 for an example of a brief written sampling procedure.

VII.

Sample Custody

Sample custody is a tracking system used as a part of good laboratory and field operations. Sample custody documents where and with whom samples are at all times from the field through laboratory measurements. The system helps in determining where a sample became contaminated or otherwise compromised so that the situation can be corrected. Where samples are or could potentially become involved in legal investigations, special tracking procedures called "chain-of-custody"[1] procedures are used. See Figure 23 for an illustration of a chain-of-custody form.

For routine (nonlitigation) samples, the considerations listed below can be addressed in a SOP[12] to provide a description of procedures, forms, documentation, and personnel responsibilities. If the sponsor's requirements allow it, this section of the QA plan can be made into a subsection of the previous

---

[1] CHAIN-OF-CUSTODY: Special tracking procedures used for samples where litigation is involved. The procedures for tracking are established by the U.S. EPA's National Enforcement Investigation Center (NEIC).[24] Full guidance including audio-visual presentations may be obtained from the NEIC concerning chain-of-custody procedures.

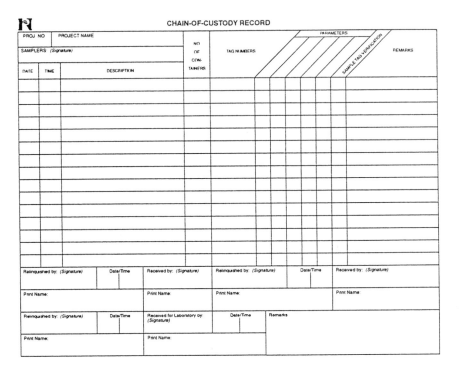

FIGURE 23.   An example of a chain-of-custody form.

section, Element VI, Site Selection and Sampling Procedures. Whenever possible reference, or include in an appendix, a written procedure for the following:

- List field and laboratory sample custodians.[4]
- List forms and notebooks and describe procedures used to record sample information.[19]
- List location, subject, sample type, conditions, and other information.
- Provide examples of forms (Figure 24), labels (Figure 25), and seals (Figure 26).[4,19]
- Describe the procedures used for completing the labels and forms.

  - Indelible ink, other considerations
  - Details of labels and forms that are not self-explanatory

CIRCLE THE APPROPRIATE RESPONSE:

1. CUSTODY SEAL	present / absent
	intact / not intact
2. CHAIN-OF-CUSTODY	present / absent
3. SAMPLE TAGS	present / absent
SAMPLE TAG NUMBERS	listed / not listed on chain of custody
4. SMO FORMS	present / absent

DATE:

BATCH NUMBER:

CASE NUMBER:

AIRBILL NUMBER:

STORAGE AREA:

DELIVERING AGENT'S SIGNATURE:

DOCUMENT CONTROL #:

				CORRESPONDING		Does information on custody records, traffic reports, and sample tags agree?	REMARKS:
Date Received	Time Received	Chain-of-Custody Record Number	Client / SMO Sample Numbers	Sample Tag Numbers	Assigned Lab Numbers		Condition of sample shipment, etc.

FIGURE 24. An example sample form used in sample control.

- Describe documentation of sample transportation including identity of the individuals responsible for transport.[4]
- Describe the procedure for transferring and maintaining custody.[19]

Harry Reid Center for Environmental Studies
University of Nevada, Las Vegas, 4505 Maryland Parkway,
Las Vegas, Nevada 89154-4009

Sample Label

Sample No. _____

Sample Type _____

Project _____

Date/Time _____

Media _____

Subject _____

Location _____

Method Preservative _____

Comments _____

_____

_____

FIGURE 25. A field sample label.

Custody Seal

Signature _____

Date _____

CUSTODY SEAL

CUSTODY SEAL

Date _____

Signature _____

FIGURE 26. A chain-of-custody seal.

VIII.

Calibration Procedures and Frequency

It is important to any research project to demonstrate that the measurement systems used in the research are operating properly. Periodic calibration[m] of the equipment and instrumentation will generally accomplish this. Many of the instruments used in life science laboratories today utilize internally automated or computerized calibrations (e.g., colony counters). Such advances in instrumentation have made it unnecessary to compose calibration tables for many measurements. However, it is important to verify the internal calibration periodically. Also for certain determinations that require a multipoint calibration to establish a concentration curve, a periodic calibration check is used to assure that the curve remains valid.

This section of the QA plan contains calibration procedures and frequencies for each measurement system, instrument,

---

[m] CALIBRATION: Comparison of a measurement standard or instrument with another standard or instrument to report or eliminate by adjustment any variation (deviation) in the accuracy of the item being compared.[1]

Measurement Parameters	Instrument Responses for 3-point Calibration				
	20	50	80	X	% RSD
A					
B					
C					
D					
E					
F					

FIGURE 27.   An example calibration table.

device, or technique used to obtain measurement data (e.g., pH meters, analytical balances, pipets, counters) that may affect the quality of the research data. It may simplify the composition of this section to present it in a summary table as in Figure 27 with references to procedures or manuals when applicable. Consider including the following information in this section of the QA plan:

- List all tools, gauges, instruments, and other sampling, measuring, and test equipment used for data collection activities that affect the quality of the research.[19]
- Describe or reference the procedures used to calibrate.
- List certified equipment and/or standards including source, traceability, and purity checks. If a nationally recognized standard (e.g., National Institute of Standards and Technology) does not exist, identify the standards used and explain how these records will be maintained and traceable to the item being calibrated.[19]
- Describe the frequency of initial calibrations, calibration checks, or recalibrations.
- Define control limits and specific acceptance criteria for all calibration measurements.[12]

IX.

Measurement Methods

Measurement methods are chosen by the research scientist as appropriate for the analytical parameters or target organisms in the specific matrix at the anticipated levels. Be sure to choose methods that will achieve the qualitative and quantitative QA objectives established for the research. Also remember that for some projects sponsors may require that only validated methods be used.

In planning the research and preparing this section of the QA plan, identify each measurement method and/or equipment needed. For each method list:

- Preparation methods
- Decontamination procedures
- Reagents/materials
- Performance requirements
- The corrective action procedure[19]
- Confirmatory measurement methods[4]

A tabular presentation may simplify composition of this section (Figure 28).

Medium	Quality Control Procedures				Organisms	Expected Reactions	Comment
	S	G	I	BR			
Brilliant green lactose bile broth	X			X	*Escherichia coli* *Staphylococcus* sp.	Positive = yellow (growth) Negative = green (no growth)	Incubation temperature 35°C, 48 hours
Azide-amended agar	X	X	X		*Streptococcus pyogenes* *Escherichia coli*	Small colony within 24 hours Growth inhibited	Azide is added to media to inhibit gram-negative organisms and allow gram positive's (*Streptococci*) to grow.

S = Sterility; G = Growth; I = Inhibition; BR = Biochemical Response

FIGURE 28. A tabular presentation of measurement procedures.

X.

Data Reduction, Verification, Validation, and Reporting

The investigator plans the project around collecting data of a quality that meets specific objectives. Therefore, it is important to assure that errors do not occur during the processes of data reduction,[n] verification,[o] validation,[p] and reporting. This

---

[n] DATA REDUCTION: The transformation of raw data into a more useful form.[25]

---

[o] DATA VERIFICATION: A routine activity conducted by technical, laboratory, and clerical personnel on small sets of the data to determine if data have been accurately quantified, recorded, and transcribed; if data have been collected and analyzed in accordance with prescribed, approved procedures; if the data appear suitably complete; and if the data appear to be reasonable and consistent, based on prior knowledge of the research.[26]

---

[p] DATA VALIDATION: The process by which a sample, measurement method, or data is deemed useful for a specified purpose;[1] and independent, timely review of a body of verified data against a predetermined set of qualitative and quantitative criteria to evaluate their adequacy for their intended use.[25]

section of the QA plan is used to organize data flow (Figure 29) and prevent the occurrence of errors. Consider the following:

- Data reduction
  This includes calculations, tabulations, and transfers that result in interpretation and presentation of data. A listing of personnel, procedures, statistical approaches, and products that result from data reduction should be described in this section including the following:

  - The individuals responsible for data reduction
  - Data reduction procedures including software or computer programs (An example of a data reduction procedure is presented in Figure 30.)
  - Statistical approach for reducing data including units and definitions of terms[4]
  - Sample data sheets[4] (Figure 31)
  - Results of blank QC measurements that will be treated in the calculations[4]
  - Preservation of calculations for estimation of uncertainty
  - Products of data reduction

- Data verification
  This determines if data have been quantified, recorded, and transcribed accurately and have been collected and analyzed in accordance with prescribed, approved procedures. Confirmation of completeness and consistency is included in this section with a listing of:

  - Individuals responsible for data verification
  - Criteria used to accept, reject, or qualify data[19]
  - Critical control points[q]
  - Forms and checklists that will be used[19]

---

[q] CRITICAL CONTROL POINT: A point where the results are so important to the study that if an indicator shows that a process or system is out of control, no further progress can be made until the problem is corrected and the control point is repeated. The potential loss of time, money, study particulars, and certain safety aspects determines how critical any control point may be.

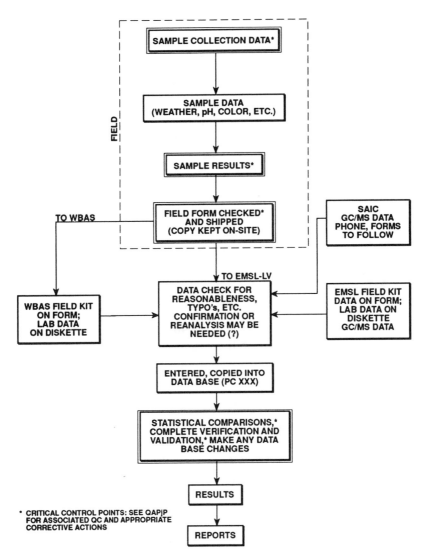

FIGURE 29. An example of a data-handling flow diagram.[13]

Example: When a calibration check (serving as a **critical control**) result is outside of acceptance criteria, the data generated since the previous calibration check are suspect, and most likely repeat analysis will be required.

1. Calculation of percent recovery for spikes of a solution with a standard of known concentration.

    a. Calculate the change in concentration Δ S expected due to the spike.

    b. Measure the concentration of the solution before spiking (C_o) and the concentration of the spiked solution (C_s).

    c. The percent recovery is the percent of the change in concentration relative to the expected change.

    $$\text{percent recovery} = 100 \ (C_s - C_o), \Delta \ S$$

2. Calculation of the coefficient of variation (% CV). The coefficient of variation is a measure of scatter or dispersion and is defined as the ratio of the standard deviation to the mean.

    $$\%CV = 100 \ (Sx / \bar{x})$$

    where $\bar{x} =$ the mean concentration for a sample of size n.
    $S_x =$ the standard deviation of the mean.

3. Calculation of accuracy, or the degree of agreement of a measurement (or a limiting mean of measurements), x, with an accepted reference or true value, t. Accuracy can be expressed as the difference between the two values, as a percentage of the reference or true value.

    $$\text{Accuracy} = 100 \ (x - t) / t$$

FIGURE 30. A portion of a data reduction procedure.[27]

- Data validation
  This determines that the data are adequate for their intended use.[26] This section discusses the personnel and activities concerned with evaluating the data and should include a listing of:

  - Individuals responsible for data validation
  - Procedures for determining outliers and flagging data (see the second half of Figure 31)
  - Criteria (DQOs) that will be used to validate data[26]

- Data reporting
  This includes the presentation of results and usually incorporated presentation of tables and illustrations with discussion and conclusions. This section of the QA plan discusses the personnel responsible for reporting the data and the form in which the report will be presented and should include:

  - Individuals responsible for data reporting
  - Format and time constraints for the data report

Laboratory Name: _____ Case No. _____

Lab Sample ID No. _____ Analytical Protocol: _____

Site: _____

Date Sample Recieved: _____

SAMPLE VALUE

Organism	CFU/g
Pseudomonas syringae	_____
Pseudomonas aeruginosa	_____
Erwinia herbicola	_____
Enterobacter agglomerans	_____

DATA REPORTING QUALIFIERS

For reporting results to EPA, the following results qualifiers are used. Additional flags or footnotes explaining results are encouraged. However, the definition of each flag must be explicit.

Value   If the result is a value greater than or equal to the detection limit, report the value.

U   Indicates organism was analyzed for but not detected. Report the minimum detection limit for the sample with the U (e.g., 1 CFU) based on necessary concentration dilution action. (This is not necessarily the detection limit.) The footnote should read U-organism was analyzed for but not detected. The number is the minimum attainable detection limit for the sample.

J   Indicates an estimated value. This flag is used either when estimating a concentration for tentatively identified organisms where the organism meets the identification criteria but the result is less than the specified detection limit but greater than zero (e.g., 1 CFU if limit of detection is 10 CFU/g and a concentration 3 CFU/g is calculated, report as 3J).

B   This flag is used when the organism is found in the blank as well as a sample. It indicates possible/probable blank contamination and warns the data user to take appropriate action.

Other   Other specific flags and footnotes may be required to properly define the results. If used they must be fully described and such description attached to the report.

FIGURE 31.   An example of a sample data sheet.

XI.

Computer Software

This section of the QA plan is used to describe any controls and processes needed to prevent defects in the research data due to computer software. The sponsor's requirements for this section can be extensive depending on what the software is being used for and whether the software was written by research personnel or purchased from a reputable manufacturer. Software used for graphics, presentations, or in other ways that do not affect research data quality does not require extensive QA. Such software is used and documented with practices consistent with its intended use.

Generally any software used to reduce research data needs to be verified to be defect-free and validated that it accomplishes its intended purpose. Certification from the manufacturer indicates that the software has been thoroughly tested and that the manufacturer stands behind the quality of its product. An additional check is for the researcher to run a known through the software to see if the software produces the correct response. If the software is used with equipment or instrumentation, this can be accomplished during standardization.

Electronic calculation programs generally do not need comprehensive QA documentation. However, software programs may need extensive QA if they involve any of the following:

- Complex manipulations
- Were written by an individual on the research staff
- Were not purchased with certification
- Are being used on a sensitive research study

QA requirements for these types of software programs may include:

- Software specification[27]

  - Program capabilities
  - Source of the data
  - Definition of valid input
  - Error checking performed on the data input

- Individuals responsible for software QA
- A description of the approach to and documentation of software

  - Development
  - Acquisition
  - Testing
  - Use

- Software life cycle[28] processes and controls[29]

  - Classification
  - Requirements development
  - Acquisition
  - Design
  - Code development
  - Installation and use
  - Modifications
  - Verification/validation
  - User application
  - Documentation and review
  - Configuration management

These requirements are based on the philosophy that software quality depends on the entire development process, not just upon testing the software product. Finally, the corrective action to be taken when software defects are found should be described in this section of the researcher's QA plan. The researcher preparing this section should remember that:

> "A major benefit of today's computer technology is that it provides the means to integrate data in ways which will lead to new scientific insights. Artificial intelligence, innovative programming, massive data storage capabilities, and vastly improved communication technology will inevitably draw diverse data sources together...the intellectual input to make this possible must come from biologists now."[15]

XII.

Quality Control Requirements

Quality control (QC)[r] requirements apply to the collection of any data that affect the quality of the research project. This includes field sampling and measurement activities. In the QA plan, the researcher lists the type and number of QC requirements including acceptance criteria and the stage at which each QC requirement will occur. This information may be presented most easily in summary tables as shown in Figures 32 and 33.

The QC requirements are dependent on the research being conducted. Basically, an investigator needs to decide which parameters, measurement systems, and sets of data must be in control[s] to attain project objectives and DQOs. The investigator then determines what type of QC can be instituted in the study to monitor each parameter or data set.

---

[r] QUALITY CONTROL: The routine application of procedures for obtaining prescribed standards of performance in the measurement process.[2]

---

[s] IN CONTROL: The state of an instrument or process when it is operating within standardized limits, usually established for a study or measurement system.

Internal QC Check Procedure	Frequency of Internal QC Checks
Temperature monitoring - incubators	daily monitoring
Bacterial content of distilled water reservoir	monthly
pH testing for glassware cleanliness	monthly
Contamination of media	each batch
Autoclave function	monthly
Accuracy of balances	monthly
Calibration of pipettors	monthly
Accuracy of pH meter	daily

FIGURE 32.   An example table of quality control checks.[5]

Criteria for weekly testing:	Minimum Inhibitory Concentration Tests (MIC)	Disk Diffusion Tests
	Document accuracy for 30 consecutive test days	Document accuracy and precision for 30 consecutive test days
Take corrective action in DAILY testing if:	Two (2) consecutive out-of-control results occur More than 2 out-of-control results occur in 20 consecutive test days	Two (2) consecutive out-of-control results occur More than 2 out-of-control results occur in 20 consecutive test days One (1) result occurs that is > 4 standard deviations ± the control limits for any zone
Take corrective action in WEEKLY testing if:	Any MIC value falls outside the accuracy control limits	Any zone value falls outside the accuracy control limits

FIGURE 33.   An example of a summary table of quality control checks.[5]

Example: If a study requires that a measurement be accurate within ±50%, then the QC reference material (discussed below) measurements should be better than ±50%.

In other words, results of the QC determinations must fall within certain limits of acceptance to support the objectives. If QC measurements do not meet acceptable limits, they become "real-time" indicators flagging unacceptable data. Immediate corrective action can then be performed preventing the collection of more unacceptable data. For some research, measurement methods may not be standardized in the initial stages of the project. Detailed QC requirements may not be possible in these cases. General QC concepts and requirements as well as commitments to pursue as many of these as possible can be put in the QA plan.[17,19]

If a published method listing QC measurements is used, the reference may be cited in the QA plan. However, existing methods often do not provide adequate QC for research projects. Therefore, the researcher needs to thoroughly review the method to assure that all critical parameters are covered and supplement QC measurements are provided where necessary.

Unfortunately, for many aspects of life science research, there are no standards or knowns to be used for QC, and often measurements are subjective. In these cases, the scientist must be creative in developing QC standards.

Example: In a study in which color determinations are made visually, an innovative QC standard for comparison is needed. Colors are subjective determinations since the results may be quite variable among personnel. However, the project leader needs assurance that the measurements are as accurate as visual determinations can be. Therefore, a QC requirement to check visual accuracy[30,31] could be the periodic use of a standardized color chart (e.g., Pantone® Matching System flip chart) or a color chart used by ornithologists.

Using color standards might not completely eliminate variability due to subjectivity, but is one way to minimize it. Another way to minimize the variability in this particular example is for the same investigator to perform all of the determinations throughout the study. The use of a color guide in this way may be considered to be a form of calibration check (discussed below) since the user is basically calibrating the eye to standardize perception of the colors observed.

Some types of QC to consider are QC samples (collocated, split, blank, or replicate) and spikes (surrogate or duplicate):

- QC samples
  - Collocated sample[t]

---

[t] COLLOCATED SAMPLE: Samples collected at the same location and time such that the two samples do not differ from one another.

Example: Two identical aerobiological samplers placed on the same sampling platform and operated at the same time. These **collocated samples** would be used to check the variability of the air samplers and/or the homogeneity of the airborne microorganisms.

    – Split sample[u]

Example: **Dividing** a sample of surface water into two containers. These samples would be used to check the homogeneity of the sample.

    – Replicate sample[v]

Example: **Repeat** enumerations of a suspension of bacterial cells with a Coulter Counter. These counts would be used to check the precision of the electronic counting device.

    – Duplicate sample

Example: Collecting **two** water samples from one randomly selected site and performing similar measurements from both samples independently.

An example of a listing of QC samples and their specific uses is presented in Figure 34.

- Spike samples
  - Surrogate spike[w]

---

[u] SPLIT SAMPLE: A replicate portion or subsample of a total sample obtained in such a manner that it does not differ from other portions of the same sample.

---

[v] REPLICATE SAMPLE: A counterpart of another sample, usually referring to an analytical sample or a measurement; the general case for which duplicate is the special case consisting of two samples or measurements.

---

[w] SURROGATE SPIKE: Quantity of a known substance similar in nature to the analyte that is not indigenous to the sample, but rather is incorporated into the sample prior to analysis.

Machine	Material	Measurement	Method	Manpower	Environment
Field Blanks	Reagent Blanks	Alternate Methods	Alternate Methods	PE Samples	Extraction Blanks
Field Duplicates	Matrix Spikes/ Duplicates	Calibration Standards	Split Samples	Proficiency Samples	Field Blanks
Standard Reference Materials	Lab Control Standards	Internal Standards	PE Samples	Lab Control Standards	Trip Blanks
Calibration Standards	Surrogates	Matrix Spikes/ Duplicates	Proficiency Samples	Blind Replicates	Field Spikes
Calibration Blanks		Standard Reference Materials	Lab Control Standards		Method Blanks
Interference Checks		Lab Control Standards			Digestion Blanks
Instrument Blanks		Method of Standard Additions			Leachate Blanks
Check Standards		Surrogates			Reagent Blanks
		Serial Dilutions			Holding Blanks

Machine	Material	Measurement	Method	Manpower	Environment
Maintenance	Homogenize	Calibration Verification	Validation	Training	Non-Contaminating Materials
Preventive Maintenance	Preservation		SOPs	Certification	Cleaning Procedures
Calibration				Readiness Review	Preservation
Cleaning					Holding Times
					Reagent Tests

FIGURE 34.   A presentation of quality control samples and their uses.[32]

Example: The use of a **control organism inoculated** into a water sample prior to assay by serial dilutions and spread plate counts. The enumeration of the control organism throughout the dilution series and spread plate counting procedure demonstrates that the samples were handled properly in the laboratory.

– Duplicate spikes[x]

Example: The inoculation of a QC spike water sample with a control organism at the field site immediately after collecting the water and inoculation of a second QC spike water sample in the laboratory. **Both spike samples** are then assayed by spread plate counts. The results of the field and laboratory spike should be similar in number for the control organism to demonstrate that transport procedures[33] do not affect the results.

– Spiked blanks

Example: The **inoculation of a sterilized water blank** with a control organism prior to assay by serial dilutions and spread plate counts. The enumeration of the control organism throughout the dilution series and spread plate counting procedure demonstrates that the samples were handled properly in the laboratory. The lack of other bacterial growth in the blank also demonstrates the sterility of handling.

– Internal standards[y]

Example: The use of solvent peak retention time in gas chromatographic fatty acid methyl ester (FAME) analysis to check on the presence of specific methylated esters.

- Other
  – Standard reference materials (SRMs)[z]

---

[x] DUPLICATE (SPIKED OR NONSPIKED) SAMPLE: A second sample randomly selected from a population of interest to assist in the evaluation of sample variance.

---

[y] INTERNAL STANDARDS: Material introduced into the sample prior to analysis that has a known, repeatable value upon analysis.

---

[z] STANDARD REFERENCE MATERIAL (SRM): A reference material distributed and certified by the National Institute of Standards and Technology, formerly of the National Bureau of Standards; a reference material is defined as a substance with one or more of its properties sufficiently well established to be used for calibration of an apparatus, the assessment of a measurement method, or for assigning values to materials.[34]

Example: The use of a **standard culture** (e.g., American Type Culture Collection strain, ATTC, Rockville, MD).

– Blanks[aa]

Example: The use of a nonexposed (air sampling) or noninoculated (water or soil sampling) agar-filled Petri dish processed through transportation, incubation, and analysis phases of a monitoring survey where microorganisms are cultured on agar-filled Petri dishes. Growth on the **blank** Petri dish would indicate contamination of the media unrelated to exposure to the test material.

Example: For a measurement protocol using instrumentation, the establishment of acceptable background counts obtained with an electronic particle counter (e.g., Coulter Counter) without any sample.

– Confirmatory measurements with a second method[4]

Example: The use of epifluorescent assay as a **second assay** to verify the identification of a *Giardia* cyst detected with phase contrast microscopy.

– Control charts[bb]

Unfortunately these charts on instruments are indicators of stability and cannot be solely relied upon to indicate accuracy or precision of measurement data.[18]

[aa] BLANK: Sample matrix without the analyte present; when an analyte is detected in a blank, it is due to artifacts; values obtained with the blank should be deducted from a measured value to give a net value that results only from the component contained in a sample.

[bb] CONTROL CHART: A graphical plot of test results with respect to time or sequence of measurement, together with limits within which they are expected to lie when the system is in a state of statistical control.

– Calibration standards[cc]

Example: The use of a mid-range pH buffer standard during pH measurements of a series of water samples. This would check that the **calibration curve** is still valid during the period when all of the measurements were recorded.

– Proficiency testing of analysts[4]

Example: The **scheduled capability testing** of laboratory technicians for their accuracy in identifying fungal cultures with known strains.

– Independent, multi-method or multi-laboratory analyses for confirmation of analytical results[4]

Example: Sending specimens to **another laboratory** for analysis with the same and different methods and comparing the results with the in-house data.

– Any other checks needed for the specific research

Example: The testing of genome stability and genetic interactions with indigenous populations during **experiments especially designed** to determine gene transfer in the environment.

For each QC check that is used, a listing of the frequency of use, the location (field or laboratory) of use, the organisms or analytes to be used, the levels, the control limits, any forms or bench sheets used (Figure 35), and the corrective actions for exceeded limits should be included.

[cc] CALIBRATION STANDARDS: A solution or substance prepared by the analyst to establish a calibration curve or the analytical response function of an instrument.

MEDIA PREPARATION QUALITY CONTROL

TYPE OF MEDIUM _____ BATCH NO. _____

PLATE _____ TUBE FLUID ☐ TUBE SOLID ☐

PREPARATION DATE _____ QUALITY CONTROL DATE _____

STUDY_____

A. QUALITY CONTROL

	acceptable	not acceptable	not done
pH	☐	☐	☐
COLOR	☐	☐	☐
CLARITY	☐	☐	☐
STERILITY	☐	☐	☐
PERFORMANCE	☐	☐	☐

SIGNED   DATE

B. USER

UNACCEPTABLE BY USER ☐   DATE USED _____

DATE REPORTED _____

IF THE MEDIUM IS UNACCEPTABLE BY THE USER, PLEASE RETURN THE COMPLETED FORM WITH
SAMPLES OF THE BATCH, ACCOMPANIED BY COMMENTS AND OBSERVATIONS TO THE MEDIA PREPARER.

COMMENTS: _____
_____

SIGNED   DATE

STUDY

C. MEDIA PREPARATION

ACTION TAKEN:_____ REPEATED ☐   DATE _____
_____   NEW BATCH NO._____
_____

PROBLEM: IDENTIFIED ☐   NOT IDENTIFIED ☐

COMMENTS:_____
_____

SIGNED   DATE

FIGURE 35.   An example of a quality control form.[18]

XIII.

Audits and Assessments

Audits or assessments are intended to reinforce to project personnel the importance of complying with the QA plan and written procedures in the field and laboratory. The assessment activity can also be used to drive the revision of these documents as well as providing an opportunity to review good laboratory practices. In this book, the authors present the types of audits and assessments as two major categories: internal and external.

Internal audits or self-assessments are performed within the defined organization. They may be internal to the investigator's program but external to the project in an organization in which there are independent QA personnel who can perform an audit function. Or they may be planned to be performed within the project by the project leader or designee.

External audits are performed by personnel external to the defined organization such as the sponsor's QA personnel or the project leader's corporate office auditors.

There are many types of audits and assessments regardless of whether they are internal or external. These include:

- Surveillance[19] — in which auditors observe the work as it is being performed
- Inspection[19] — in which items are inspected for compliance

- Peer review[19] — in which peers of the scientific discipline review a document such as a report or study plan for technical merit
- Readiness review[19] — in which an operation is evaluated prior to initiation
- Technical system audit[19] — in which a system is evaluated against criteria
- Performance evaluation — in which a system, operation, or work group is evaluated based on actual performance often against a performance evaluation sample
- Data quality assessment[19] — in which a set of data is evaluated against specific criteria

Audit and assessment activities can be designed to be a review of routine data or performance sample data only or may entail visits to the laboratory or field for an on-site evaluation. Most audits are based on some degree of compliance. Personnel who perform assessments and audits (auditor) may audit to a general checksheet that applies to similar laboratory or field operations (Figure 36) or perhaps strictly to the research study's QA plan and procedures. Sponsors have different philosophies, and the research may even be audited on compliance to the sponsor's QA requirements. Often audits range widely in degrees of formality. They may be informal in which the auditors exceed the checksheets they are using and make helpful suggestions regarding laboratory or field procedures. Or the audits may be so formal that nothing beyond finding written evidence that plans and procedures are complied with is addressed by the auditor.

The following items should be addressed in this section of the QA plan:

- Type of audit or assessment including goals of the audit
- Schedule of planned audits
- Personnel responsible for audits or assessments and their relation to the group being audited
- Any training or certification that is a requirement of the audit team[29]
- Discussion of how corrective response actions will be handled based on the audit or assessment findings and how the actions will be verified[29] and recorded

Document: MicroAudit.lst
Revision Number: 2
Date: May 13, 1993
Page: 13 of 26

10.2	**CENTRIFUGE ROUTINE MAINTENANCE**
10.2.1.	QUARTERLY SPEED CHECKS
	Yes ☐ No ☐ Comments _____
	_____
10.2.2.	QUARTERLY CHECK AND CALIBRATION OF VARIABLE SPEED DIAL
	Yes ☐ No ☐ Comments _____
	_____
10.2.3.	QUARTERLY TIMER CHECKS
	Yes ☐ No ☐ Comments _____
	_____
10.2.4.	INSPECTION AND CLEANING OF:
	Yes ☐ No ☐ Comments _____
	_____
10.2.4.1	CHAMBER
	Yes ☐ No ☐ Comments _____
	_____
10.2.4.2.	HEAD
	Yes ☐ No ☐ Comments _____
	_____
10.2.5.	SUPPORT SHAFT BEARINGS LUBRICATED AT 3 – 6 MONTH INTERVAL
	Yes ☐ No ☐ Comments _____
	_____
10.2.6.	SHAFT CHECKED FOR SIDE PLAY IN THE BEARINGS, QUARTERLY
	Yes ☐ No ☐ Comments _____
	_____
10.2.7.	BRUSHES CHECKED FOR WEAR EVERY 3 – 4 MONTHS
	Yes ☐ No ☐ Comments _____
	_____
10.2.8.	COMMUTATOR CHECKED FOR SCRATCHES AND DIRT
	Yes ☐ No ☐ Comments _____
	_____

Figure 36.   A page from an audit checklist.[18]

Since many scientists have not experienced an audit, it may be useful for the project leader and/or QA staff to prepare the study personnel. It should be made clear to all that an audit is an assessment or review and not a police procedure. A description, with as much detail as possible, should be given as to what the audit will involve. It may be useful to prepare personnel for an external audit or assessment by performing a self-assessment. Providing project personnel with the audit checksheet, if available, is also helpful.

XIV.

Preventive Maintenance

Life science researchers are using more instrumentation in the laboratory, and it is often critical to the study that this equipment is operational and reliable. This section includes a brief description of types of preventive maintenance needed for adhering to research schedules and for completing objectives. Although seemingly unimportant to many investigators, it can be useful to clarify exactly where each scientist's responsibility for instruments and equipment begins and ends. Stating the division of responsibility for maintenance is often important in a setting where there are service contracts on instruments. This is also helpful where funding is limited.

The actual procedures for preventive maintenance are best addressed by reference to instrument/ equipment procedures. These are often the manufacturer-supplied manuals and instructions.

In addition to the following listed topics, any important issues that the investigator has encountered previously that can be prevented are included in this section:

- A schedule of preventive maintenance — examples are shown in Figures 37 and 38.

81

Ministry Laboratory of Services Health Branch Ontario	Equipment Quality Control Centrifuge		MOH No.			
			Location			

Date	Clean All[4] Spillage	Clean[1] Interior	Clean[1] Exterior	Brush[2]		Check[3] RPM	Tech Initials
				Check	Change[4]		

1755-44 (87/08)   [1]perform monthly   [2]perform quarterly   [3]perform semi-annually   [4]perform as necessary

FIGURE 37. An example of a preventive maintenance form.[18]

- An inventory of critical spare parts and supplies (e.g., media, reagents, Petri dishes, glass slides) that must be kept in-house and who is responsible for this supply. Lack of these items can cause a major research delay.

Instrument					Identification Number					

Phone # for Service					Date of Purchase					

Quarterly	D	P	D	P	D	P	D	P	D	P
Item 1										
Item 2										
Item 3										

D - Date Due
P - Date Performed

Semiannual	D	P	D	P
Item 4				
Item 5				
Item 6				

Remarks: Repairs, Parts, Adjusts, Down Time

FIGURE 38.   An example of a preventive maintenance schedule.[5]

- Maintenance contract information for critical measurement systems including current phone numbers, account numbers, and other information needed to facilitate service. These may be placed in procedures and/or manuals.
- Archive policy and location of one-of-a-kind manuals and instructions including all versions of computer programs used to reduce and analyze data.
- Discuss how inspections and acceptance testing is to be performed and recorded.[19]
- Training of instrument and equipment operators.
- Describe continuity in the case of breakdowns.
- Specify recordkeeping requirements, such as what, who, when, and where this is recorded.

XV.

Calculation of Data Quality Indicators

It is important to plan in detail the data assessment procedures, especially statistical treatment, including equations, units, and assessment frequency to be used in a project. This section of the QA plan must agree with the DQOs in order to be effective. Parameters that are particularly critical are accuracy, precision, completeness, comparability, and method detection limit. The following formulas (Figures 39 to 42) are examples only. Other formulas[35] exist and may be more project-specific.

> *Example:* In the formula for **precision,** by always requiring C_1 to correspond to instrument number 1, one can detect a systematic change from a time series of RPD.

Some projects will require that data be assessed for completeness, representativeness, and comparability. See Figures 43 and 44. See QA Element V for an explanation of comparability and representativeness. Many studies use a critical measurement system requiring the calculation of detection limits (Figure 45).

$$RPD = \frac{(C_1 - C_2) \times 100}{(C_1 + C_2)/2} \quad ;$$

RPD = relative percent difference,
C_1 = larger of the two observed values, and
C_2 = smaller of the two observed values.

FIGURE 39. A formula for the calculation of precision from duplicate measurements.[4]

$$RSD = (s/\bar{y}) \times 100 \quad ;$$

RSD = relative standard deviation,
s = standard deviation, and
$\bar{y}$ = mean of replicate analyses.

FIGURE 40. A formula for the calculation of precision from replicates using relative standard deviation.[4]

$$s = \sqrt{\sum_{i=1}^{n} \frac{(y_i - \bar{y})^2}{n-1}} \quad ;$$

s = standard deviation,
y_i = measured value of the i-th replicate,
$\bar{y}$ = mean of replicate measurements, and
n = number of replicates.

FIGURE 41. A formula for the calculation of accuracy from the true value of a spike.[4]

$$\%R = 100 \times \left[\frac{C_m}{C_{srm}}\right] \; ;$$

$\%R$ = percent recovery,

C_m = measured concentration of SRM, and

C_{srm} = actual concentration of SRM.

FIGURE 42. A formula for the calculation of accuracy from the true value of a standard reference material.[4]

$$\%C = 100 \times \left[\frac{V}{T}\right] \; ;$$

$\%C$ = percent completeness,

V = number of measurements judged valid, and

T = total number of measurements.

FIGURE 43. A formula for the calculation of sampling and analytical completeness.[4]

$$\%C = 100 \ \times \left[\frac{V}{n}\right] \ ;$$

$\%C$ = percent completeness,
V = number of measurements judged valid, and
n = total number of measurements necessary to achieve a specified statistical level of confidence in decision making.

FIGURE 44. A formula for the calculation of statistical completeness.[4]

$$MDL = t_{(n-1, 1-\alpha = 0.99)} \times S \quad ;$$

MDL = method detection limit,
S = standard deviation of the replicate measurements, and
$t_{(n-1, 1-\alpha = 0.99)}$ = students' t-value appropriate to a 99% confidence level and a standard deviation estimate with $n - 1$ degrees of freedom.

FIGURE 45. A formula for the calculation of method detection limits.[4]

XVI.

Corrective Action

It is rarely possible to predict every problem that can occur in a study. However, for certain procedures that are performed, the investigator has previously encountered or is aware of some potential problems. For the predictable routine problems, corrective actions are normally addressed within the written standard operating procedures to save time in preparing the QA plan. However, this information can be placed wherever it is most useful to personnel. This section of the QA plan specifies corrective action for predictable nonroutine problems such as results of interlaboratory comparison studies[12] or not meeting DQOs. Corrective actions as a result of audits and assessments may be addressed as response actions in the audits or assessments section of the QA plan.

It is possible for the investigator to design corrective actions with IF...THEN... statements[5] (Figure 46). With these kinds of statements, the investigator can specify to personnel the action to take when a predictable nonroutine problem occurs.

For the more elusive failures such as may occur in the previously untried portions of the study or the totally unexpected failures in routine procedures, corrective actions are more difficult to devise. However, for each determination critical to the success of the project, the following topics can be addressed:

"During weekly testing, if a zone or minimum inhibitory concentration (MIC) value is observed outside the established accuracy control limits, then the following control checks are required:

1. Appropriate control strain(s) must be tested for 5 consecutive days.

2. For each drug-microorganism combination, all of the above five zones or MIC values must be within the accuracy control limits.

3. In the disk diffusion test, zone range of the five consecutive tests for each drug-microorganism must be within the precision range described earlier.

If any result in either item 2 or 3 above is out of control, the laboratory must resume daily control testing for a minimum of 20 consecutive test days and, therefore, must again meet the accuracy and precision requirements before attempting to return to weekly testing."

FIGURE 46.   An example of a corrective action statement.[5]

INCIDENT REPORT - SAMPLE REJECTION					
Date	Reported to:	Sample ID	Reason Sample Rejected	Action	Tech's Initials

FIGURE 47.   An example of a corrective action report.[5]

- Trigger points

  Define pre-specified conditions that will automatically require corrective action.

- List personnel

  Specify who will initiate, approve, implement, evaluate, and record corrective action (Figure 47).

- Describe response

  List specific procedures to use when corrective action is indicated.

XVII.

QA/QC Reports to Management

This section of the QA plan helps the project leader specify the individuals who will prepare reports, the content of reports, and communicate data effectively to management. Describe the frequency and distribution of reports prepared. Topics to consider for inclusion in QA reports to management are:

- Type of report

  - Written or oral
  - Interim or final

- Report contents

  - Changes in plans and procedures
  - Training
  - Results of audits and assessments
  - Significant problems, recommended solutions, and results of corrective actions
  - Summary of data quality assessments
  - Discussion of whether the DQOs are being met
  - Limitations on use of the measurement data, and discussion of the effects of such limitations on the defensibility of the data

Glossary

Accuracy:
The degree of agreement of measurements with an accepted reference or true value. Accuracy is usually expressed as the difference between the two values, or the difference as a percentage of the reference or true value, and sometimes is expressed as a ratio.

Blank:
Sample matrix without the analyte present; when an analyte is detected in a blank, it is due to artifacts; values obtained with the blank should be deducted from a measured value to give a net value that results only from the component contained in a sample.

Calibration:
Comparison of a measurement standard or instrument with another standard or instrument to report or eliminate by adjustment any variation (deviation) in the accuracy of the item being compared.[1]

Calibration standards:
A solution or substance prepared by the analyst to establish a calibration curve or the analytical response function of an instrument.

Chain-of-custody:
Special tracking procedures used for samples where litigation is involved. The procedures for tracking are established by the U.S. EPA's National Enforcement Investigation Center (NEIC).[24] Full guidance including audio-visual presentations may be obtained from the NEIC concerning chain-of-custody procedures.

Collocated sample:
Samples collected at the same location and time such that the two samples do not differ from one another.

Comparability:
The degree to which one data set can be compared to another. It is not expressed arithmetically. Comparability is achieved by using consistent methods and traceable standards.[12]

Completeness:
A measure of the amount of valid data obtained from a measurement system compared to the amount expected to be obtained under correct normal conditions.

Control chart:
A graphical plot of test results with respect to time or sequence of measurement, together with limits within which they are expected to lie when the system is in a state of statistical control.

Critical control points:
A point where the results are so important to the study that if an indicator shows that a process or system is out of control, no further progress can be made until the problem is corrected and the control point is repeated. The potential loss of time, money, study particulars, and certain safety aspects determines how critical any control point may be.

Data quality:
Features and characteristics of data that bear on their ability to satisfy a given purpose. The characteristics of major importance are accuracy, precision, completeness, representativeness, and comparability.[2] Data quality can also be expanded to include timeliness and value to the client.

Data quality objectives (DQOs):
Statements describing the level of uncertainty a decision maker is willing to accept in results derived from environmental data. DQOs are defined when the results are going to be used in a regulatory or programmatic decision.[9]

Data reduction:
The transformation of raw data into a more useful form.[25]

Data validation:
The process by which a sample, measurement method, or a piece of data is deemed useful for a specified purpose;[1] an independent, timely review of a body of verified data against a predetermined set of qualitative and quantitative criteria to evaluate their adequacy for their intended use.[25]

Data verification:
A routine activity conducted by technical, laboratory, and clerical personnel on small sets of the data to determine if data have been accurately quantified, recorded, and transcribed; if data have been collected and analyzed in accordance with prescribed, approved procedures; if the data appear suitably complete; and if the data appear to be reasonable and consistent, based on prior knowledge of the research.[26]

Duplicate sample:
A second sample randomly selected from a population of interest to assist in the evaluation of sample variance.

In control:
The state of an instrument or process when it is operating within standardized limits, usually established for a study or measurement system.

Internal standards:
Material introduced into the sample prior to analysis that has a known, repeatable value upon analysis.

Matrix spikes:
Quantity of a known substance similar in nature to the analyte that is not indigenous to the sample that is incorporated into the sample matrix prior to analysis.

Precision:
A measure of mutual agreement among individual measurements of the same property, usually under prescribed similar conditions. Precision is best expressed in terms of the standard deviation. Various measures or precision exist depending upon the "prescribed similar conditions."

Quality assurance (QA):
The total integrated management program for assuring the reliability of data. A system for integrating the quality planning, quality assessment, and quality improvement efforts to meet user requirements.

Quality assurance project plan (QAP):
An orderly assemblage of detailed and specific procedures that delineates how data of known and accepted quality are produced for a specific project or study.

Quality control (QC):
The routine application of procedures for obtaining prescribed standards of performance in the measurement process.[2]

Quality management plan (QMP):
An orderly assemblage of management policies, objectives, principles, and general procedures by which an agency or laboratory outlines how it intends to produce data of known and accepted quality.

Replicate sample:
A counterpart of another sample, usually referring to an analytical sample or a measurement; the general case for which duplicate is the special case consisting of two samples or measurements.

Split sample:
A replicate portion or subsample of a total sample obtained in such a manner that it is not believed to differ from other portions of the same sample.

Standard operating procedure (SOP):
A written document that details an operation, analysis, or action whose mechanisms are thoroughly prescribed and that is commonly accepted as the method for performing certain routine or repetitive tasks.

Standard reference material:
A reference material distributed and certified by the National Institute of Standards and Technology, formerly the National Bureau of Standards; a reference material is defined as a substance with one or more of its properties sufficiently well established to be used for calibration of an apparatus, the assessment of a measurement method, or for assigning values to materials.[34]

Surrogate spike:
Quantity of a known substance similar in nature to the analyte that is not indigenous to the sample, but rather is incorporated into the sample prior to analysis.

Type I and Type II errors:
Expressions of the likelihood that a false positive or a false negative will affect the study hypothesis.[22]

References

1. Taylor, J. T., *Quality Assurance of Chemical Measurements*, Lewis Publishers, Chelsea, MI, 1987, Appendix A.

2. U.S. Environmental Protection Agency, *Interim Guidelines and Specifications for Preparing Quality Assurance Project Plans*, QAMS-005/80, EPA-600/4-83-004, U.S. EPA, Washington, D.C., 1983.

3. U.S. Environmental Protection Agency, *OTS Guidance Document for the Preparation of Quality Assurance Project Plans*, U.S. EPA, Office of Toxic Substances, Washington, D.C., 1987.

4. Simes, G., *Preparing Perfect Project Plans*, EPA/600/9-89/087, U.S. EPA, Risk Reduction Engineering Laboratory, Cincinnati, OH, 1989.

5. Miller, J. M., *Quality Control in Microbiology*, U.S. Dept. of Health and Human Services, U.S. Public Health Service, Centers for Disease Control, Atlanta, GA, 1987.

6. Clinical laboratories improvement act of 1967, *Federal Register* 33, October 15, 1968.

7. Clesceri, L. S., Greenberg, A. E., and Trussell, R. R., Eds., *Standard Methods for the Examination of Water and Wastewater*, 17th ed., American Public Health Association, American Water Works Association, and Water Pollution Control Federation, Washington, D.C., 1989.

8. Kanare, H. M., *Writing the Laboratory Notebook*, American Chemical Society, Washington, D.C., 1985, chap. 7.

9. U.S. Environmental Protection Agency, *Development of Data Quality Objectives: Description of Stages I and II*, unpublished report, U.S. EPA, Quality Assurance Management Staff, Office of Modeling, Monitoring Systems & Quality Assurance, Washington, D.C., 1986.

10. U.S. Environmental Protection Agency, *Quality Assurance Program for Biotechnology Team*, U.S. EPA, Environmental Research Laboratory, Corvallis, OR, 1990.

11. U.S. Environmental Protection Agency, *Planning for Environmental Research Using the Data Quality Objectives Process*, draft, U.S. EPA, Quality Assurance Management Staff, Washington, D.C., 1991.

12. Simes, G., *Preparation Aids for the Development of Category I - IV Quality Assurance Project Plans*, EPA/600/8-91/003-006, U.S. EPA, Risk Reduction Engineering Laboratory, Cincinnati, OH, 1991.

13. U.S. Environmental Protection Agency, *Superfund Innovative Technology Evaluation Program Demonstration Plan for Westinghouse Bio-Analytic Systems Pentachlorophenol Immunoassays*, U.S. EPA, Environmental Monitoring Systems Laboratory, Las Vegas, NV, 1990.

14. Erdmann, D. E., *Technical Review of Water-Quality Laboratories Providing Analytical Services for the Water Resources Division of the U.S. Geological Survey*, open-file Report 91-223, Denver, CO, 1991.

15. International Council of Scientific Unions (ICSU), Problems of access to standardized biological terminology, Committee on Data for Science and Technology, in *CODATA/Newsletter* 56, 4, Paris, 1991.

16. U.S. Environmental Protection Agency, *Quality Assurance Glossary and Acronyms*, unpublished report, U.S. EPA, Quality Assurance Management Staff, Office of Modeling, Monitoring Systems & Quality Assurance, Washington, D.C., 1991.

17. Poziomek, E. J. and Cross-Smiecinski, A. J., Quality assurance plans for basic research and new concept studies, in *Proceedings of the Ninth Annual Waste Testing And Quality Assurance Symposium*, U.S. Environmental Protection Agency, Arlington, VA, 1993.

18. Association of Analytical Chemists, International, *Quality Assurance for Microbiological Laboratories Short Course Manual*, Assoc. Analytical Chemists, Internat., Cincinnati, OH, September 3–4, 1992.

19. U.S. Environmental Protection Agency, *EPA Requirement for Quality Assurance Project Plans for Environmental Data Operations*, revised draft, EPA QA/R-5, U.S. EPA, Quality Assurance Management Staff, Research Triangle Park, NC, 1993.

20. Donegan, K., Matyac, C., Seidler, R. A., and Porteous, A., Evaluation of methods for sampling, recovery, and enumeration of bacteria applied to the phylloplane, *Appl. Environ. Microbiol.*, 57, 51, 1991.

21. Murphy, R. B., On the meaning of precision and accuracy, Paper 6.1 in National Bureau of Standards Special Publication 300, 1969, Vol. 1.

22. Sokal, R. R. and Rohlf, F. J., *Biometry*, W.H. Freeman, San Francisco, 1981, 159.

23. U.S. Environmental Protection Agency, *Soil Sampling Quality Assurance User's Guide*, U.S. EPA, Environmental Monitoring Systems Laboratory, Las Vegas, NV, EPA/600/8-89/046, 1989.

24. U.S. Environmental Protection Agency, *NEIC Policies and Procedures*, EPA-330/9-78-001-R, U.S. EPA, Office of Enforcement and Compliance Monitoring, National Enforcement Investigations Center, Denver, CO, 1986.

25. Parker, S. P., *McGraw-Hill Dictionary of Scientific and Technical Terms*, 4th ed., McGraw-Hill, New York, 1989.

26. Ramos, S. J., A project manager's primer on data validation, in *Proceedings Nineteenth Annual National Energy & Environmental Division Conference*, American Society for Quality Control, Milwaukee, WI, 1992.

27. Morris, C. R., Computer validation, *Network News*, International Center for Health and Environmental Education, Alexandria, VA, 7, 1, 1993.

28. Harris, S. D., Software quality assurance requirements of a nuclear program written to meet current software quality assurance standards, in *Proceedings Nineteenth Annual National Energy & Environmental Division Conference*, American Society for Quality Control, Milwaukee, WI, 1992.

29. U.S. Department of Energy, *Quality Assurance Requirements Document*, DOE/RW-0214, Revision 4, U.S. DOE, Office of Civilian Radioactive Waste Management, Washington, D.C., 1990.
30. Smithe, F. B., *Naturalist's Color Guide*, American Museum of Natural History, Washington, D.C., 1975.
31. Smithe, F. B., *Naturalist's Color Guide*, American Museum of Natural History, Washington, D.C., 1974.
32. Mills, P. E., Take a representative sample, in *Proceedings Nineteenth Annual National Energy & Environmental Division Conference*, American Society for Quality Control, Milwaukee, WI, 1992.
33. Khlebnikova, G. M., Gilichinskii, D. A., Fedorov-Davydov, D. G., and Vorob'eva, E. A., Quantitative evaluation of microorganisms in permafrost deposits and buried soils, *Soil Microbiology*, Plenum, 1990, 59, 105.
34. National Institute of Standards and Technology, *NIST Standard Reference Materials Catalog 1990–1991*, U.S. Department of Commerce, Gaithersburg, MD, 1990.
35. U.S. Environmental Protection Agency, *A Rationale for the Assessment of Errors in the Sampling of Soils*, EPA/600/X-89/203, U.S. EPA, Environmental Monitoring Systems Laboratory, Las Vegas, NV, 1993.

Index

acceptance criteria, 54
acceptance ranges, 40
accuracy, 36, 37, 85, 93
  assessment of, 14
  establishing, 38
  visual display of concept of, 39
aerobiological samplers, 70
aerobiological study, 41
archive policy, 83
audit checklist, page from, 79
audits, assessments and, 77–79

biological research projects, 2
blanks, 73, 93
breakdowns, 83

calibration, 53, 93
  procedures, 53–54
  standards, 74, 94
  table, example, 54
certifications, 31
chain-of-custody, 94
  form, 50
  procedures, 47, 49
  seal, 52
charts, 27
chemical measurements, 45
Clinical Laboratory Improvement Act, 2
clinical laboratory testing, 2

collocated sample, 69, 70, 94
color standards, 69
communication numbers, 32
communications, mechanisms of, 31–32
comparability, 36, 40, 85, 94
completeness, 36–38, 85, 94
  calculation of, 87
  confirmation of, 58
component activities, 28
computer software, 63–65
computer technology, major benefit of, 65
confirmatory measurements, 55, 73
conflict of interest, 29
control chart, 73, 94
controlled document, 8
control limits, 54
control organism, 71
corrective action, 89–90
  procedure, 55
  report, example of, 90
  statement, example of, 90
critical control, 59
critical control points, 58, 94
custody, 45, 52

data
  bases, comparison of, 41
  collection, 5, 34, 54, 57
  flow, organization of, 58

quality
  assessment, 78
  indicators, calculation of, 85–88
  objectives (DQOs), 5, 6, 33, 34, 89, 95
  reduction, 57–61, 95
    procedures, 58, 60
    products of, 58
  reporting, 57–61
  set, monitoring of, 67
  validation, 57–61, 95
  verification, 57–61, 95
decontamination procedures, 55
detection limits, calculation of, 85
device, calibration procedures for, 54
document, 8
  control, 7–10
    agreement form, 8
    list, 8, 10
    plate, 7, 9
    record, 8, 11
    stamp, 8, 9
  number, 12
DQOs, *see* data quality objectives
duplicate measurements, 86
duplicate sample, 70, 95
duplicate spikes, 72

electronic calculation programs, 63
error, 34, 57
experimental design, 27
external audits, 77

false null hypothesis, acceptance of, 43
false positive result, 43
field operations, 49
field sample label, 52
field sampling, 67
flow diagrams, 27, 59
format preference, 23
formatting, 6

general overview, 27

holding times, of samples, 47
hypothesis, 27, 34, 43, 44

in control, 14, 67, 95
inspection, 77
instrument
  calibration procedures for, 53
  operators, training of, 83

instrumentation, 14, 81
internal audits, 77
internal standards, 72, 96

life science research, 45
  applying QA to, 2
  QC standards, 69
locator page, 23, 25

maintenance contract information, 83
management organizational chart, 30
management structure, 29
matrix spikes, 96
measurement
  levels, 46
  methods, 55–56
  objectives, 33
  parameters, 47
  procedures, tabular presentation of, 56
  quality objectives, 34, 42, 43
  system, 35, 53
method detection limits, 42, 85, 88
microbiology, 2
modified method, 37, 38
multi-laboratory analysis, 74
multi-organization study, responsibilities, 31
multiple measurement methods, 42

National Institute of Standards and Technology, 54
nonlitigation samples, 49
nonquantitative objectives, 34
Nuclear Regulatory Commission, 7

peer review, 78
performance evaluation, 78
performance requirements, 55
precision, 36, 37, 85, 96
  establishing, 38
  visual display of concept of, 39
preliminary studies, 28
preparation methods, 55
preventive maintenance, 81–83
  form, example of, 82
  schedule, 81, 83
project
  activities, timeline of, 28
  description, 27–28
  leader, 29
  -level document, 3

management, 37
objectives, 34, 35
organization, 29–32
personnel, 79

QA, *see* quality assurance
QAP, *see* quality assurance project plan
QC, *see* quality control
QMP, *see* quality management plan
quality assurance (QA), 1, 96
  documents, hierarchy of, 4
  project plan (QAP), 3–6, 96
    elements, 13–15
    compliance with, 77
    composition of, 2
    concurrence with, 19
    controlled status of, 7
    formatting of, 6
    preparation of, 45
    revisions of, 10–12
    table of contents for, 23
    types of, 3
  requirements, 15
  staff, 13, 19
  terminology, 33
quality control (QC), 1, 67, 96
  check, example table of, 68
  form, example of, 75
  measurements, blank, 58
  procedures, description of, 47
  requirements, 67–75
  samples, uses, 71
  standards, developing, 69
quality management plan (QMP), 3, 96
quality objectives, 33–44
  concept of, 33, 37
  determining, 37–38
  impact of not meeting, 43
  types of, 34
  value of, 34
quantitative objectives, 38–40

readiness review, 78
reagents, 55
real-time indicators, 68
recalibrations, frequency of, 54
recordkeeping, 45, 83
replicate sample, 70, 97
report
  contents, 91
  types of, 91

representativeness, 36, 41
research
  data
    prevention of defects in, 63
    software used to reduce, 63
    validity of, 1
  measurement systems used in, 53
  objectives of, 29
  organization, 19, 29
  personnel, 13
    structural organization of, 29
    tasks of, 30
  study, quality objectives used in, 36
  supporting information for, 12
revision number, 8
revision summary, 10, 11
routine task, description of, 4

sample
  control, 45, 51
  custody, 49–52
  data sheet, 58, 61
  history, recording, 47
  matrix, 73
  transportation, documentation of, 52
  types, 46
  volumes, 47
sampling equipment, 47
sampling frequencies, 46
sampling plans, 13
sampling procedures, 47
sampling site, location of, 46
scientific investigation, sequence of steps
    in, 6
scientific notebooks, 4, 5
scientific objectives, 34
signature requirements, 22
sign-off sheets, 8
site modifications, 46
site selection, sampling procedures and,
    45–48
small study, organization chart for, 32
software
  certification from manufacturer, 63
  defects, 64
  documentation of, 64
  life cycle, 64
  programs, QA requirements for, 64
  QA, 64
  specification, 64
SOP, *see* standard operating procedure

spare parts, inventory of, 82
specificity, 42
spike
  sample, 70
  true value of, 86
spiked blanks, 72
split sample, 70, 97
sponsor, 19
  approval, 21
  requirements, 34, 49, 78
SRMs, *see* standard reference materials
standard culture, 73
standard operating procedure (SOP), 3, 48,
  97
standard reference materials (SRMs), 72,
  87, 97
standards, identification of, 54
statistical support, 45
statistical completeness, calculation of, 88
storage, 45
study situations, conceptual model of, 5
subtasks, determining of, 29
summary tables, 27, 40, 42, 68

supplementary quality objectives, 42
surrogate spike, 70, 97
surveillance, 77

table of contents, 23–25
tables, 27
target organisms, 46
tasks, determining of, 29
technical QA/QC roles, 29
technical system audit, 78
timelines, 27
title page, 19–22
Type I and Type II errors, 15, 43, 44, 97

U.S. EPA, 3
  in-house project, 22
  research study, 20
  quality objectives of, 33
  QA plan requirement, 37
  reporting of data, 41

water blank, inoculation of sterilized, 72